T3-AMD-947

IS ANYONE THERE?

ISAAC ASIMOV

SF

ace books

A Division of Charter Communications Inc.

A GROSSET & DUNLAP COMPANY

51 Madison Avenue

New York, New York 10010

IS ANYONE THERE?

To John M. Blugerman
pleasant brother,
warm-hearted uncle

This Ace printing: August 1980

Printed in U.S.A.

Contents:

CONTENTS

PART II CONCERNING THE MORE OR LESS UNKNOWN:

A—OTHER LIFE

CONCERNING THE MORE OR LESS UNKNOWN:

B—FUTURE LIFE

CONTENTS

INTRODUCTION

I am known to be a science fiction writer. I am also known to be a member of the faculty of Boston University School of Medicine. Consequently, I suppose it is natural that I am often asked what my colleagues think of the fact that I write science fiction.

Perhaps the questioner has the feeling that I probably meet with disapproval; that my path is strewn with tacks over which I must walk barefoot; that my professional career is blunted and stultified.

It is rather a disappointment to me to have to deny the drama, but my professional life is not a hard one. Some of my colleagues are unaware that I write science fiction, and wouldn't know how to care less if they did know. Others are aware of it but consider it as just another academic idiosyncrasy. Still others happen to be science fiction readers and often read my stories—I hope with pleasure. And a few, by Heaven, are science fiction writers themselves.

This is not to say that there wasn't a time when I myself wondered whether an academic career and a reputation as a wild-eyed science fiction writer might prove to be incompatible.

The possibility struck me most forcibly in June 1949, when two things happened. First, I was about to join the medical school faculty. Second I had just sold my first science fiction novel to Doubleday & Co., and it was to appear as a "regular book."

I had been selling science fiction stories to the magazines of the genre for eleven years before that, to be sure, but I always felt that to be an obscure exercise that was a secret between myself and an odd scattering of exotic Fans. But a *book* was a different thing; it couldn't be kept secret, could it?

Fortunately, I was caught in no quandary, trapped in no dilemma, haunted by no uncertainty. From an early age, I had known I was a writer, and I had also known that if I ever had to choose between writing and something else, I would always choose writing. (Knowing one's course of action in advance lends one a great peace of mind and it is to this I attribute my freedom from ulcers despite a way of life that is made up almost exclusively of deadlines.)

Hence I saw no need to act irresolutely. If I was ever to be faced with a choice, let it be *now*. So I made an appointment with the dean.

I said to him politely, but firmly, "Sir, as you know, I am a new instructor in biochemistry here. However, I feel it only fair to tell you that in a few months my first science fiction novel will be published as a book and the medical school will find itself indirectly identified with it."

To which he asked in his turn, "Is it a good book?"

I answered, cautiously, "Doubleday thinks so."

"Well, then," he said, "we'll be glad to be identified with it."

And that was it. In the years that have since passed, no one at school has objected to my science fiction to my face; and no one has brought me any report of any objection behind my back either.

Another crisis arose in my mind when I began to publish non-fiction. In 1952, I was part-author of a textbook of biochemistry for medical students, and since then I have published many non-fiction books on a wide variety of subjects.

At the start, I wondered whether I might not be plagued to use a pseudonym. "Come, come, Asimov," I heard some ghostly editor say in my ear, "we can't ruin the sales of a serious book by having prospective readers say: 'This can't be good; that nutty science fiction writer wrote it.' "

I was prepared for Homeric battles, for I was determined

to have my name on everything I wrote. (In the first place, I like my name; in the second place, I am self-centered; in the third place, I am proud of science fiction and of my place in it and I won't have it insulted.)

The Homeric battle, alas, never took place. No editor—not one—ever objected to the science fictional halo that rests slantedly over my amiable head. In fact, I began to notice that in many cases, the little biographical squib placed on the flaps of the book jackets of my most serious non-fiction took care to mention my science fiction as proof of the fact that I could write well.

That drove me back to the final bastion of possible lack of appreciation, the mass media. Good science fiction, after all, appealed to the minority; there was no getting round that. Outlets that, of necessity, had to appeal to a broad and miscellaneous readership had to give it a miss.

That carefully reasoned conclusion broke into little pieces with the coming of the space age in 1957. Suddenly the most mass and the most medium of the mass media were interested in the oddest matters. They began to desire articles concerning matters on the frontiers of science, and they even grew increasingly interested in science fiction. (Last year, *The Saturday Evening Post* published a science fiction novel of mine—something I never thought would happen.)

Again I found that my science fiction background did not hinder; rather, it helped. I was asked to write articles by publications I would scarcely have dared approach a few years earlier. Pretending nonchalance, I wrote the articles and soon found that, while I retain my post on the faculty, I had to give up teaching. I am a full-time writer now.

How different the situation is from what it was as late as 1949. Then I was convinced I was working in total obscurity and that if I were to ask the question "Is Anyone There?" with respect to my audience, the answer would come back from a vast echo-sounding emptiness: "No one but us science fiction fans, Asimov."

But now as I look about the rather large array of miscellaneous writings for which I am responsible (all of them based, to begin with, on my reputation as a science fiction writer) I can ask the question again, and know that the answer will be clamorously many-tongued.

And to come full circle, here we are at Doubleday again, where my first novel was published. The gentlemen here seem perfectly willing to publish a collection of my articles from the general magazines, suitably revised and updated wherever necessary. Some of these articles deal with science, some with speculation, and some with science fiction—the three legs of my tripod.

PART I

CONCERNING THE MORE OR LESS KNOWN

A
LIFE

Chapter 1   Matter Over Mind

What is mind? No matter!
What is matter? Never mind!

This ancient witticism testifies to man's firm conviction through the ages that the human mind transcends the material, that it is not bound by the ordinary rules that govern gross matter.

The physical structure of the living organism is accepted as a thing of atoms and molecules, governed by the same laws that govern the rocks underfoot and the stars overhead. That is as true for Man the Proud as for Worm the Lowly. But man's mind? Can one analyze the creative genius that gives rise to a masterpiece? Can one weigh, count, and measure emotion and imagination, love and hate, passion, thought, and a sense of good and evil?

There has always been a strong impulse to place mind over matter and to apply different and more subtle rules to the former. It seems natural, then, that doctor's medicine should prove unable to work on the mind. Shakespeare has Macbeth ask cynically of the doctor treating his nightmare-ridden wife:

> *Canst thou not minister to a mind diseased,*
> *Pluck from the memory a rooted sorrow,*
> *Raze out the written troubles of the brain,*
> *And with some sweet oblivious antidote*
> *Cleanse the stuffed bosom of that perilous stuff*
> *Which weighs upon the heart?*

This appeared under the title "That Odd Chemical Complex, The Human Mind" in *The New York Times Magazine*, July 3, 1966.

To which the doctor can only reply humbly:

> *Therein the patient*
> *Must minister to himself.*

Three centuries after Shakespeare, when doctors began to "minister to a mind diseased," they did it without any "sweet oblivious antidote," without any potion, nostrum, or material device. To reach the mind the laws of matter were insufficient; the mind itself had to be the tool. Doctors began to talk to patients and, more important, to listen while patients talked. In place of the physician's stethoscope and the clinician's test tube we had the psychiatrist's couch.

Physical scientists have been strongly tempted to leave it at that and to make no move toward the mentally disturbed person upon the psychiatrist's couch. To approach the vast complexities of the mind with the cold, material instruments of science required a kind of heroism. There was a grim promise of inevitable failure about the fire-breathing dragon of mind-chemistry that tended to daunt the would-be St. George of the microscope and the slide rule.

And yet the brain is made up of atoms and molecules—as is the rest of the body. The molecules in the cells of the body, and in those of the brain in particular, are so many and so various and so versatile that they interact and change in a dazzling pattern that we do not, as yet, understand well. But the very dazzle of this chemical complexity is hopeful, for it is, quite conceivably, complex enough to account for all the nearly infinite subtlety of what we call the mind.

This complexity is now being tackled by new techniques that are making top news out of advances in brain chemistry and physiology. Computers are being used to analyze brain-wave data with a completeness never before possible. Greater understanding of nucleic acids in connection with the machinery of heredity is producing exciting hints concerning the mechanics of memory (something I shall take up in some detail in Chapter 2).

Most of all, new drugs are being used that affect the workings of the brain, sometimes drastically, and by that very fact are giving us possible insights into these workings.

16

It is this last technique that has been creating the most stir, for it involves, among other things, the compound called LSD, which offers mankind a new dimension in drug use and drug consequence.

The new advances, striking as they do at the most subtle manifestations of the brain—memory, perception, reason— do not come from nowhere. There is a century of advance in connection with the less complex aspects of brain action. Although the nervous system is an intricately interlaced whole on almost every level of its activity, it shows, in some respects, a sort of gradually increasing complexity of function from the bottom upward. This has helped scientists move onward by easy stages until now they can reasonably try to cope with the mental machinery that knits together all levels of the nervous system.

Below the brain is the spinal cord, a narrow 18-inch-long mass of nerve tissue that runs down the center of the bones making up the spinal column. The spinal cord is a switching center for many of our common reflexes. Touch something hot and the sensation sparks its way to the cord and is converted into an outward-surging nerve impulse calling for a quick withdrawal. Your finger moves away even before your conscious mind has a chance to say, "It's hot."

(Mind you, this is not to say that this is *all* the spinal cord does. It is knit by nerve tracts to the various centers of the brain and it forms part of a unified whole. However, it is this reflex action that was first understood, and I am deliberately over-simplifying to get across the historical perspective.)

At its upper end the spinal cord widens into the medulla oblongata, or "brain stem," upon which the brain itself sits like a swollen piece of wrinkled fruit. The brain stem handles matters that are more complicated than the simple reflexes. It is an important center for the control of the manner in which we stand, for instance.

In standing, we are actively using muscles to keep our back and legs stiff against the pull of gravity. To do this efficiently, there must be a constant, delicately adjusted interplay. No one set of muscles is allowed to overbalance us to one side or another without a countering set being quickly thrown in to readjust the balance accurately. We are not

17

ordinarily aware of this activity, but if we have been standing a long time, weariness makes itself unpleasantly evident, and if we lose consciousness while standing, the muscles relax and we crumple to the ground at once.

If it were our conscious mind that were continually concerned with the muscles involved in standing, we would have little time for anything else. It is the brain stem that is in charge, however, with scarcely any conscious interference. We remain standing, balancing ourselves accurately, no matter how distracted we are, no matter how lost in thought, provided only that we are not actually asleep or unconscious.

Above the brain stem are two swollen bodies with wrinkled surfaces, each divided nearly in half. The larger is the cerebrum (the Latin word for brain); the smaller, in the rear, is the cerebellum (little brain).

The cerebellum goes one step beyond the brain stem. It does more than keep us balanced while motionless; it keeps us balanced in motion. While we walk, we lift one leg, throw ourselves off balance temporarily and then move the leg forward and bring it to a halt upon the ground in just the manner calculated to retrieve that balance. If we move our hand toward a pencil, that hand must begin to slow before it reaches the pencil and must come to a halt just as it reaches it.

There must be "feedback." We must see (or otherwise sense) the motion of a portion of our body, estimate its distance from its goal and adjust its speed and direction constantly on the basis of the changing situation. It is the cerebellum that is in charge of this. It takes care of the matter automatically, so that if we reach for a pencil we seize it with perfect efficiency, without any awareness of the delicacy of the task. But watch someone with cerebral palsy who cannot manage this feedback. He is unable to perform the slightest task without a pathetic overshooting and undershooting of the mark.

In accomplishing all this, incoming sensations must produce chemical changes in the brain cells which, in turn, give rise to nerve impulses that produce specific muscle responses. What the details of these chemical changes might be we don't know.

As we come to the cerebrum we find ourselves more

directly involved with chemistry. At the bottom of the cerebrum, for instance, is a section called the hypothalamus, one of the functions of which is to act as a thermostat. The body's heat is produced through a constant gentle vibration of the muscles at a rate of from seven to thirteen times per second, a fact reported in 1962. The hypothalamus senses the temperature of the blood passing through it. If that temperature is too low, it sparks an increase in the vibration rate, producing additional heat. If the temperature is too high, the vibration rate is lowered. This is one way in which body heat is maintained at an almost constant level despite changes in outside temperature.

The hypothalamus also detects the water concentration in the blood and acts through a nearby gland, the pituitary, to adjust the workings of the kidney accordingly. More water is eliminated if the blood is getting too thin; less water, if it is too thick. The hypothalamus is also constantly measuring the sugar concentration in the blood. When that concentration falls too low, the hypothalamus acts to set up hunger sensations (see Chapter 3).

Here we have clearer examples of actual chemical involvement. Small (as yet harmless) chemical changes in the blood call forth alterations in the body's mechanism to prevent any further (and increasingly harmful) changes in that direction. The body's chemistry is thus kept in accurate balance.

The details must be extraordinarily complex, however. The body's mechanism is intricately interconnected, and the hypothalamus must bring about desirable changes in one part of that supercomplicated network without bringing about undesirable changes elsewhere. The difficulty here is exemplified by the manner in which almost every man-applied drug, despite the most careful use, has always the possibility of bringing on unpleasant "side effects." The hypothalamus must work with the kind of incredible sure-footedness that avoids side effects.

But what about the upper parts of the cerebrum—the parts particularly concerned with conscious motion and sensations, with thought and reason, memory and imagination? If we are stumped by the chemistry of such things as reflexes and water balance, surely we must be completely,

19

hopelessly, and helplessly at sea in connection with the chemistry of memory, for instance?

In fact (as we shall see in the next chapter), we are not. We are actually making progress, or seem to be, in the understanding of memory, and the most exciting prospects may be looming on the far horizon.

And it is not only the reasonably healthy mind that is in question. What we call mental disorders may simply be shifts in the chemical workings of the brain. If mental disease is a material malfunction, then through the study of brain chemistry we may well find the cures that have steadily eluded the psychiatrists.

Consider schizophrenia, for instance—the most common of the serious mental illnesses. The name was coined in 1911 by a Swiss psychiatrist, Paul E. Bleuler, from the Greek words meaning "split mind" because it was frequently noted that persons suffering from this disease seemed to be dominated by one set of ideas (or "complex") to the exclusion of others, as though the mind's harmonious workings had been disrupted and split, with one portion of that split mind seizing control of the rest. An older name for the disease was dementia praecox ("early ripening madness"), a term intended to differentiate it from senile dementia, mental illness affecting the old through the deterioration of the brain with age. Schizophrenia shows itself at a comparatively early age, generally between 18 and 28.

Schizophrenia may exist in several varieties, depending on which complex predominates. It may be hebephrenic ("childish mind"), where one prominent symptom is childish or silly behavior. It may be catatonic ("toning down"), in which behavior is indeed toned down and the patient seems to withdraw from participation in the objective world, becoming mute and rigid. It may also be paranoid ("madness"), characterized by extreme hostility and suspicion, with feelings of persecution.

At least half of all patients in mental hospitals are schizophrenics of these or other types, and it is estimated that one percent of mankind is affected. This means that there are at least 30 million schizophrenics in the world, a figure equal to the total population of a nation like Spain.

Can this most common variety of the mind diseased be treated by "some sweet oblivious antidote"?

There are precedents that give us ground for hope. Some mental illnesses have already been cured, and the mind has shown itself amenable to physical treatment—in certain cases at least.

One example is pellagra, a disease once very common in Mediterranean lands and in our own South. It was characterized by what were called the three D's: diarrhea, dermatitis, and dementia. As it turned out, pellagra was caused by a vitamin deficiency, the lack of niacin in the diet. Once niacin was supplied to patients in the necessary quantities the disease cleared up. Not only did the diarrhea stop, not only was the red, inflamed, roughened skin restored to normal, but the mental disorders ceased. The same chemical that healed the body healed the mind. In this instance, at least, it was a case of matter over mind.

Pellagra is caused by a failure of supplies from outside. But what about malfunctions caused by inadequacies in the body's own chemical machinery? Every chemical reaction in the body is controlled by complex substances known as enzymes; each reaction has its own particular enzyme. What, then, if a person is born without the ability to manufacture some particular enzyme?

This is the situation in cases of a disease called phenylpyruvic oligophrenia, which is characterized by serious mental deficiency. This disease (not common, fortunately) is present at birth. A child is born without the ability to make a certain enzyme that brings about the conversion of a substance called phenylalanine into another called tyrosine. The phenylalanine, unable to follow its normal course, changes into other, abnormal substances. These abnormal substances accumulate and interfere with brain chemistry.

Here, unfortunately, the situation cannot be corrected as simply as in the case of pellagra. Although it is easy to supply a missing vitamin, it is as yet impossible to supply a missing enzyme. However, some improvement in mental condition has been reported among patients with the disease who have been kept on a diet low in phenylalanine.

Is it possible, then, that schizophrenia is also the result of a chemical failure, either from without or within? Dr. A. Hoffer at the University Hospital in Saskatoon, Canada, has been treating schizophrenia for years by the administration of large doses of niacin and has been reporting considerable success. Apparently at least some forms of schizophrenia are a vitamin-deficiency disease rather like a more serious pellagra.

It takes more niacin to handle schizophrenia than pellagra, and Hoffer suggests a reason for this. Niacin is converted in the body into a more complex substance called NAD, which is what really does the work. The normal body can form NAD out of niacin easily and quickly if the latter is present in the diet. (Hence pellagra is cured as soon as small quantities of niacin are added to the otherwise deficient diet.) But the schizophrenic may have a disordered chemistry, characterized in part by the inability to form NAD easily. Therefore, a great deal of niacin must be supplied as a means of seeing to it that the inefficient chemical machinery produces at least a little NAD.

Hoffer reports that in the first half of 1966 he tried administration of NAD, ready-made, with very hopeful results. Smaller doses produced more rapid improvement. (As is usual in the case of experimental treatments on the border of the known, there have also been reports from other laboratories that NAD treatment has proven disappointing.)

The chemical failure in the case of the schizophrenic (whether it is the inability to make NAD out of niacin, or something else altogether) is apparently something that is inherited; for certainly the tendency to develop the disease is inherited. The chance of a particular individual in the general population developing schizophrenia is about 1 in 100. If, however, a person has a brother or sister who is schizophrenic, he has a one in seven chance of becoming schizophrenic himself. If he has an identical twin who is schizophrenic, his own chances rise to three in four.

People aren't usually born with schizophrenic symptoms to be sure; it is not inborn in the sense that phenylpyruvic oligophrenia is. We might put it this way: The schizophrenic is born not with a part of his chemical machinery missing but rather with a part that is fragile and wears out relatively early in life. It is the fragility of the part that is inherited.

But what is it that NAD (if it is NAD) does that keeps the body normal? What goes wrong in the body if NAD is missing?

Suspicion has fallen upon a portion of the chemical scheme that begins with a substance called adrenalin. In very tiny quantities, adrenalin stimulates certain nerves controlling the heart beat, blood pressure, breathing rate and so on. The adrenal gland (a small bit of tissue over each kidney) has, as one of its functions, the secretion of adrenalin into the blood stream in times of stress. When we are angry or afraid, adrenalin is produced at once so that our blood pressure rises, our heart beats faster, our lungs pump more rapidly. We are placed on an emergency footing that fits us to fight or run.

Naturally, it is important that, once the emergency is over, the body be returned to normal. For that reason the body has chemical devices for the rapid destruction of adrenalin. This destruction is supervised by an enzyme called amine oxidase, which combines with adrenalin and holds it still, so to speak, while it is altered into harmlessness.

But suppose the enzyme is occupied in some other direction? Ordinarily, enzymes are quite specific, that is, they will deal only with certain molecules possessing one particular shape and will not work with any others. This is the "lock-and-key" view of enzyme-workings (see Chapter 7). A particular key will open a particular lock, and other keys will not do.

Enzyme specificity is not perfect, however. An enzyme may combine with a molecule that is nearly but not quite the shape of the right one. The wrong molecule then competes with the right one for union with the enzyme, and if the enzyme is busy with the wrong molecule it cannot work with the right one, so that its action is inhibited. This phenomenon is called "competitive inhibition" and it can be serious indeed.

When the enzyme unites with the right molecule, it performs a task upon it and lets go; but when it unites with the wrong one it may find itself more or less permanently stuck, like a wrong key jammed into a lock and broken off there. When that happens, even a tiny quantity of a wrong molecule can bring about a long-continued chemical disorder that may do permanent damage or even bring about death. Poisons generally work in this way.

Perhaps, then, some enzyme, amine oxidase or some other, is subjected to competitive inhibition by something that is formed in the absence of NAD but not in its presence.

The possibility that competitive inhibition is indeed involved is pointed up dramatically by the case of a cactus, native to the American Southwest, that contains the compound called mescaline. The mescaline molecule bears a certain general resemblance to adrenalin—apparently close enough to allow mescaline to interfere with amine oxidase. This kind of interference, even with a single enzyme, can have a widespread effect upon brain function. The chemical workings of the brain can be likened to a vast three-dimensional lacework, intricately interconnected. A jab or yank at any one portion is going to move and shift every other portion to one extent or another. Consequently, when the portions of the cactus containing mescaline are chewed, the adrenalin-destroying enzyme is occupied with the mescaline and the adrenalin accumulates, producing all sorts of effects. A person experiences sense perceptions that have no objective existence. Ordinary objects take on strange and bizarre overtones. In short, the mescaline produces hallucinations and is therefore a "hallucinogen."

Furthermore, the reactions of the mescaline eater are inappropriate to the real universe. They depend on his distorted sense perception—and sometimes don't even match those. His behavior becomes peculiar and unpredictable. The Indians of the Southwest, experiencing all this when they ate the cactus, made the rather natural assumption that they were opening a door into a world beyond the common one of the ordinary senses. They made use of mescaline, therefore, in religious rites.

Mescaline-induced behavior resembles that of schizophrenics, and it is natural to wonder if perhaps a chemical may be formed within the body which produces effects similar to those of mescaline. Perhaps the chemical is formed more easily in the case of NAD deficiency, so that people born with a tendency to develop inefficiencies in the NAD-manufacturing reactions will therefore be subjected to the effect of these chemicals.

In the test tube, adrenalin can be easily altered to a slightly changed compound called adrenochrome. Adrenochrome, if injected into the blood stream, will also produce temporary bouts of schizophrenic-like behavior. To be sure, adrenochrome isn't formed in the normal body, but it might be in the schizophrenic.

It became a matter of interest, then, to study and analyze in detail those portions of the schizophrenic body which could be easily obtained and tested—the blood, for instance, or the urine. Any substance that could be found in all, or almost all, schizophrenics and was not found in all, or almost all, nonschizophrenics would be instantly suspect.

One way of testing body fluids is to use a technique called paper chromatography. Different kinds of molecules in the fluids are made to spread out and occupy separate spots on pieces of porous paper. These spots can then be made visible by allowing the molecules occupying them to undergo a chemical reaction that produces a colored material.

In 1962, Arnold J. Friedhoff of New York University found that with a certain course of treatment a pink spot could be obtained from the urine of 15 out of 19 schizophrenics, but from not one of 14 nonschizophrenics.

Similar tests have since been conducted on larger numbers of people. In one series of experiments, conducted by C. A. Clarke at the University of Liverpool, not one pink spot was found in tests on 265 healthy people—or on 126 people who were sick with diseases other than schizophrenia. Pink spots *were* found, however, with 46 out of 84 schizophrenics. Most of the schizophrenics who did not show the pink spot were of the paranoid variety. Among the nonparanoids, four out of every five showed it.

And what was that pink spot? It turned out to be a chemical called dimethoxyphenylethylamine (DMPE), and its structure lies somewhere between adrenalin and mescaline!

In other words, certain schizophrenics (whether for lack of NAD or some other cause) *form their own hallucinogens* and are, in effect, on a permanent mescaline kick.

This is only a bare beginning in the physical-chemical attack on schizophrenia, but it is a hopeful beginning. The pink spot (and any other chemical giveaways that may turn up) can help doctors spot the oncoming of schizophrenia earlier than might otherwise be possible and at a time when therapy might be easier. By studying the chemical processes that give rise to the pink spot, the abnormal section of the chemical mechanism in a schizophrenic may be detected and treatment might then be sharpened.

But adrenalin is not the only chemical that seems to be intimately concerned with the workings of the brain. There is also a substance called serotonin.

Serotonin's importance was brought out most dramatically in connection with lysergic acid diethylamide, the now-famous LSD. LSD has a structure somewhat more complicated than serotonin, but chemists can easily trace out a serotonin "backbone" in the LSD molecule. It is not surprising, then, that LSD may compete with serotonin for a particular enzyme as DMPE competes with adrenalin—and with the same results. In other words, the ingestion of LSD may lead to the accumulation of serotonin in the brain and the appearance of schizophrenic-like symptoms.

This fact was discovered accidentally in 1943, when a chemist, Dr. Albert Hofmann, was working on LSD with some perfectly ordinary chemical purpose in mind. He must have gotten a few crystals on his fingertips and transferred them to his lips, for he fell into a dreamlike state that left him unable to work. He returned home and experienced a kind of drunken fantasy of hallucination. He suspected it was the LSD and the next day (with remarkable courage) he swallowed about a hundred-thousandth of an ounce of it, risking only what he thought was a small test dose. It was actually a large dose, as it happened, for a tenth of that quantity would have been sufficient. The fantasies and hallucinations returned and the rest is history.

Hofmann was completely normal after 24 hours, and he did no harm to himself or to others while he was under its influence. Unfortunately, this is not something we can rely on as a general rule. Each individual has a chemical machinery of his own, so that the effect of LSD will vary from one person to another. One will experience a mild case of fantasy, others a severe one; some will recover quickly, others much more slowly.

The chemical machinery is, in some individuals, more fragile at particular key points than in others in the sense that it may be more prone to snap at those points. If the point in question is one which would produce schizophrenia if broken, taking LSD is certainly not an advisable experiment.

Ordinarily, the fragile point in the chemical scheme might hold up quite well through a long lifetime of ordinary stress so that a person might be schizophrenia-prone, without ever developing schizophrenia. Yet under the powerful jab of LSD, the point gives, and what might be merely an unusual and temporary experience for someone else becomes a permanent and serious change in the man in question.

Since none of us know just how firm some crucial part of our chemical fabric might be, using LSD without the greatest of professional care is a kind of mental Russian roulette. It is an invitation to temporary insanity for all—and possibly permanent insanity for some.

LSD is an important tool for research into mental illness. It is by studying the causes of illness that we may work out the cure. We can see that from the medical researchers who, a century ago, were led to study dangerous bacteria in order to work out a cure for infectious disease. By and large they succeeded and it is to be hoped that the second half of the 20th century will be to mental disease what the second half of the 19th was to infectious disease.

But there is one important difference. College students in the late 19th century didn't think it was exciting fun to inject themselves with cholera bacilli.

Chapter 2   *I Remember, I Remember*

It is a common thing to equate a good memory with great intelligence. The quiz programs of a decade ago were widely considered to be rewarding genius when they were actually paying off on trick memories that were sometimes

This appeared under the title "Pills to Help Us Remember?" in *The New York Times Magazine*, October 9, 1966.

(not always, of course) quite unrelated to the actual criteria of a high-powered mind.

An extreme example that came to light in this connection was the case of twins who made the newspaper headlines by their ability to give the day of the week for any date given them, through thousands of years into the past; and to do it quickly and correctly.

How they do this is not known. Have they simply memorized the calendar, or some weekday-finding summary of the calendar? Do they know the day of the week for key dates and calculate quickly from these landmarks? It is impossible to say. Nor can the twins help, for they can't explain at all. They are mentally retarded.

What's more, their freak ability does not carry over into any other branch of calculation. Even simple additions and subtractions are beyond them.

Such prodigies are by no means unheard of in history. An eighteenth century Englishman, Jedediah Buxton, could multiply 23,145,789 by 5,642,732 by 54,965 in his head and quickly get the correct answer, but he was of dull mentality just the same and remained a day laborer all his life. Zerah Colburn, born in Vermont in 1804, could give the answer to 8^{16} (sixteen 8's multiplied together) in a few seconds and work out the cube root of 268,336,125 almost at once. Nevertheless, he was not remarkably bright.

There are a number of other cases of the sort. How do they do it? It is probably a matter of a nearly indelible memory for numbers. The calculations they do in their heads can be done by anyone of ordinary intelligence on paper, since then one can write down partial products and other intermediate steps. The calculating prodigy "writes down" the numbers and the partial products in his brain and can "see" them there. There are cases of prodigies who can work their way half through a problem, pass on to other things, and then, after long periods, return to the abandoned problem, picking up where they left off without trouble. If they can do such problems with amazing speed, that may result from constant, intense, and single-minded practice.

To be sure it is not necessary to be of only normal intelligence or less in order to be a calculating prodigy. Truly great mathematicians like André M. Ampère, John Wallis,

Leonhard Euler, and, greatest of all, Carl Friedrich Gauss, had prodigious memories. These memories, however, while they aided the mathematicians in their work, were not the cause of their genius.

Still, if we leave out of account the prodigies, whether supernormal or subnormal in intelligence, we find that, in general, memory and intelligence march hand in hand. The brighter a person, the better his memory. The size of the vocabulary one understands and uses is, for instance, a pretty good indication of both the efficiency of one's memory and the extent of one's intelligence.

If we were to ask, then, what made one person have a better memory than another, we could only say that whatever it was, it was the same unknown physical cause that made one person more intelligent than another.

Theories of memory, whether ancient or modern, seem to involve one of two possibilities: memory by association or memory by image. Almost all of us accept these theories as a matter of course. We tie a knot on our finger to remember to buy bread; each time we happen to notice the knot we say, "Oh, yes; buy bread." After we've done that a few times, the matter is firmly fixed in our mind. The association has become a kind of image.

The Russian physiologist, Ivan P. Pavlov, managed to establish "conditioned reflexes" in animals by means of continued associations. He rang a bell, then showed a dog food and the dog responded by salivating. Eventually, after a number of repetitions, the dog associated the bell with food so strongly that it salivated at the sound of the bell alone. The dog's salivating mechanism had come to "remember" that the bell meant food.

This led to a school of psychology called "behaviorism" which, in its most extreme form, held that all learning and all responses were the result of conditional reflexes. It was as though you remembered a poem by heart, because you associated each phrase with the one before; or because each phrase stimulated the next phrase as a conditional response.

Yet there is no question but that memory is not necessarily merely a sequence of cause and response, of one thing reminding you of another which in turn reminds you of still another and so on. One *can* remember in images.

If I may use myself as an example (I know my own memory best) I have but an indifferent memory for numbers. I cannot multiply two three-digit numbers in my head without a great deal of trouble. However, I have a clear map of the United States imprinted in my mind and I can look at it and copy off the names of all the states as fast as I can write. (When I was young, I used to win nickles by betting I could write down the names of all the states in less than five minutes.)

Memory also comes in duration-varieties. There are short-term memories and long-term memories. If you look up a phone number, it is not difficult to remember it until you have dialed; it is then automatically forgotten. A telephone number you use frequently, however, enters the long-term memory category. Even after a lapse of months, you can dredge it up.

It is easy to suppose that a memory starts short-term and becomes long-term with use. To see what I mean, let's consider the structure of the nervous system.

The nervous system is made up of numerous microscopic cells called neurons. These are irregular in shape, with fine projections jutting out in this direction or that. These projections are called dendrites from a Latin word for tree because they resemble the branches of a tree. A particularly long process called an axon may be inches or even feet long. The dendrites or axon of one neutron may approach other neutrons very closely, but they do not quite touch. The tiny gap that remains is a synapse.

A neuron, when stimulated, is capable of transmitting a tiny electric current along its surface and down its various projections. Ordinarily, the current might stop at a synapse, but under certain conditions, the chemical environment at the synapse changes in such a fashion as to allow the current to jump the gap and pass through another cell. By jumping one synapse or another, an electric current can follow some specific path from any part of the nervous system to any other.

Suppose, then, that with every sensation you receive, a particular group of synapses is somehow affected in such a way as to make passage of the nerve-current easier. The group of synapses is so chosen that the current flows from one cell to another to another and finally it keeps on going over and over that cycle for a period of time, like auto racers lapping a track. The original sensation and a particular current-cycle can be viewed as associated. As long as the body can somehow sense a particular current-cycle and select it from all others (how such "recall" takes place, no one yet knows), it can remember the sensation that set up that particular current-cycle.

With time, however, the effect on the synapses wears off, the current-cycle fades away, and the memory is gone. It has been a short-term memory.

But each time the current-cycle is somehow sensed and the memory recalled, it may be that the change in the synapses is intensified, so that the current becomes stronger. Eventually, even the physical structure of the cells may be changed; more dendrites may form between the cells making up the cycle, thus easing the way for the current. Eventually, the current may be so firmly set that it will continue indefinitely without additional reactivation. The memory has become long-term.

Naturally, the longer a current-cycle has been in existence, the more firmly it has a chance to set, and for many of us it is therefore considerably easier to remember items learned as a youngster, than other items learned last year.

Perhaps in some exceptional cases, brains are so constructed that certain types of long-term memories, such as those involving numbers, form with particular ease, giving rise to prodigies even where the brain is not so constructed as to impart intelligence as well. Perhaps some types of current-cycles are, through usage, more easily formed and more easily set than others, so that you have a person who can remember names but not faces, or the absent-minded professor who has a viselike memory for all aspects of his subject but has difficulty recalling his home address.

But is there room enough for all the different current-cycles in the brain? Some estimates are that the brain, in a lifetime, absorbs as many as one quadrillion—1,000,000,-000.000,000—separate bits of information.

There are some ten billion (10,000,000,000) gray cells or neurons in the brain and about nine times as many auxiliary glial cells. (Some have suggested that the small glial cells are involved in short-term memory while the larger neurons handle the long-term.) If each current-cycle involves only two cells, then there is room for ten million quadrillion cycles —room for ten million times as many memories as could conceivably be accumulated in a lifetime. Of course, there are large numbers of cells that are not neighboring, but on the other hand, current-cycles can involve many more than two cells, dozens if necessary. If dozens of cells are involved, then there is more than ample room for all the current-cycles we would need.

It may even be that the brain has not only plenty of room for the necessary cycles, but has ample room to set up each cycle in many copies, for quite extensive surgery can be performed on the brain without serious impairment of the memory function. If some copies of individual cycles are removed, so to speak, by surgery, other copies remain in the parts of the brain left intact.

And yet can we be sure of something so seemingly obvious as the fact that short-term memories become long-term memories? Sometimes, when portions of the brain are stimulated electrically (for certain legitimate reasons during operations) a flow of memory results. This flow is filled with such realistic detail that the patient virtually relives a portion of his past life in full even while he simultaneously remains fully conscious of the present. Wilder G. Penfield, at McGill University, could, in this manner, cause a patient, at will, to hear snatches of music and experience scenes of childhood.

Findings such as these tempt one to suppose that the brain contains a perfect and indelible impression of all the sensations it receives. All memories are long-term, it would seem, but are quickly blocked off unless this is prevented by repeated recall. (In that case, prodigies might suffer from an imperfect blocking mechanism.)

To Sigmund Freud and his followers, such blockage of memory is by no means automatic or mechanical. Rather, it involves an active process, albeit an unconscious one. Individual memories may be chosen to be forgotten for some reason; because they are painful, embarrassing, shocking, because they brought punishment or humiliation, because they don't fit a chosen scheme of life. The process is one of "repression."

The repression isn't perfect, and some analysts suggest that neurosis is the result of the very imperfection of the act of forgetting. That which the mind would like to forget will bob up inconveniently and must then be masked, often in an irrational (that is, a neurotic) way. The cure of the neurosis may then depend, according to Freudian thinking, on dragging the memory into the open through free association, dream analysis, or other techniques. Once the memory is in clear view, it can be dealt with rationally, rather than neurotically.

Not all psychiatrists are of the Freudian school, however, and surely one might argue that forgetting can't always be a matter of harmful repression. If the brain is a perfect memory instrument, selective forgetting is necessary to survival. If you remembered every telephone number you had ever seen or heard, how difficult it would be to place your finger upon the important number you wanted among all the trivial ones you would never want.

In fact, what is the recall mechanism? Even after allowing for selective forgetting, much remains. How do you select the one item from a possibly large group of similar items in your mind?

Or, to be personal again, I am rather glib on historical names and dates. Ask me when Queen Elizabeth I died and I will answer 1603 without perceptible pause, and say 336 B.C. just as quickly if asked when Philip of Macedon was assassinated. How I select those dates so easily I don't know. I can detect no perceptible effort and am aware of no particular system.

The difficulties of determining where it is in the brain that current-cycles of memory might be concentrated, of trying to follow them once located, and, indeed, of discovering whether they exist at all are surely problems of the first magnitude. Can one shift the attack to another area instead —from physiology and cells, to chemistry and molecules, perhaps? As long ago as 1874, the English biologist, T. H. Huxley, had suggested that there was a separate key molecule in the brain for each separate memory.

The move from cells, which can at least be seen, to molecules, which cannot, might seem to be from the terribly difficult to the flatly insuperable, but it is not. Rather it resembles the story of the doctor who told his patient with a bad cold to drench himself with water and sit in a draft. "But, doctor," protested the patient, "that will turn my cold into pneumonia." "Exactly," said the doctor, "and *that* we can cure."

By the 1950s, biochemists had become increasingly confident that a certain intricate compound called ribonucleic acid (usually abbreviated RNA) was involved in the manufacture of protein. This fit in well with earlier discoveries to the effect that RNA was present in high concentration in just those cells which manufactured unusually large quantities of proteins. These included cells that were growing and multiplying and also cells that produced quantities of protein-rich secretions.

Oddly enough, however, the cell that was richest in RNA was the brain cell, and yet brain cells neither grew, multiplied, nor produced secretions. Why all the RNA then?

A Swedish neurologist, Holger Hyden, tackled this problem at the University of Gothenburg. He developed techniques that could separate single cells from the brain and then analyze them for RNA content. He took to subjecting rats to conditions where they were forced to learn new skills—that of balancing on a wire for long periods of time, for instance. By 1959, he had discovered that the brain cells of rats that were forced to learn increased their RNA content to a point where it was 12 percent higher than that of the brain cells of rats allowed to go their normal way.

RNA is thus implicated in learning and, therefore, in memory (without which learning is impossible). But is this conceivable? Granted that a set of a hundred billion cells could include current-cycles in sufficient numbers to include a lifetime of memories—how could one squeeze them all into the structure of a single molecule?

Well, the molecule of RNA is made up of a long string of four closely related, but distinctly different, units. Each item in that chain can be any one of the four units: A, B, C, or D. Two neighboring units could be any of 4 X 4 or sixteen different two-unit combinations: AA, AB, AC, AD, BA, BB, BC, BD, CA, CB, CC, CD, DA, DB, DC, or DD. Three neighboring units can be any of 4 X 4 X 4, or sixty-four different combinations, and so on.

The possible number of different combinations builds up at a tremendous rate. An RNA molecule made up of merely 25 units can have any of one quadrillion different combinations, if each unit of the molecule can be any one of the four different kinds. This means that if every different sensation experienced by a human being in the course of a long lifetime were somehow "filed away" in his brain as a different RNA combination-of-units, a 25-unit molecule would be sufficient for the task.

But RNA molecules contain many hundreds of units and not merely twenty-five. There is no question, then, but that the RNA molecule represents a filing system perfectly capable of handling any load of learning and memory that the human being is likely to put upon it—and a billion times more than that quantity, too.

Suppose that we picture a kind of "RNA memory." All cells can manufacture RNA molecules quickly and easily, but ordinary cells can prepare them in a limited variety only, to do certain limited tasks. What if brain cells can prepare them in limitless sets of combinations? Every different sensation might cause, somehow, the production of a slightly different RNA molecule. The use of that molecule at any future time might, somehow, bring back the associated sensation as a memory.

And, to be sure, Hyden found that the RNA in his learning-stimulated rats altered in nature as well as increased in kind. The ratios of the four different units were changed, as though the rats, in forming many new combinations, made use of the different units in proportions different from those they ordinarily required.

How does a brain cell go about responding to a sensation by forming an RNA molecule? Does it form any combination at random and is that combination then "assigned" to the memory of the particular sensation that brought about its formation? If that were the case, might not an RNA molecule be formed that has already been formed on another occasion, and might not memories be confused? The answer to this last question is: Probably not. The number of possible combinations is so great that the chance of accidental duplication is virtually zero.

Yet there is also the possibility that the RNA combination

35

for a particular sensation is fixed; that the particular sensation would give rise to the same RNA molecule in any creature.

A possible choice between these alternatives arose out of the work of James V. McConnell at the University of Michigan. He experimented with flatworms (planaria) about an inch and a half long. He subjected them to a flash of light and then to an electric shock. Their bodies contracted at the shock and eventually began to contract as soon as the light shone, even when the shock did not follow. They had become conditioned; that is, they had "learned" that the light meant a coming shock, and presumably they had formed new RNA molecules to take care of that bit of learning and memory.

Such trained planaria were chopped up and fed to untrained planaria which were then subjected to the same process. In 1961, McConnell reported that untrained planaria which ate the trained ones learned to react to the light faster than ordinary planaria did. They had incorporated the new RNA molecules from their food and had "eaten memory," one might say.

This meant that a particular RNA molecule was somehow tied to a particular sensation. The molecular combinations could not be selected at random, since the RNA molecule formed by planarium 1 in response to certain sensations "made sense" to planarium 2.

Allan L. Jacobson, who had worked with McConnell, continued such experiments at the University of California. If one planaria eats another, it is hard to tell which molecule in the food is being used. Why not, then, extract RNA from conditioned planaria and inject only that into unconditioned ones. That worked too. The conditioning was injected, after a fashion, along with the RNA.

And why restrict matters to planaria? (Some research workers insisted that the response of planaria was so difficult to observe that one could not be certain which ones were conditioned or whether any were conditioned at all.) Jacobson conditioned rats and hamsters to respond to the sound of a click or the flash of a light by going to a feeding box. Once conditioned, they were killed and RNA from their brains was injected into animals that hadn't been conditioned. The animals receiving these injections were found to be easier to train for they already had some of the necessary RNA they would be required to form. Interestingly, the injection worked across species. A rat could benefit by injections of RNA from a hamster.

When McConnell's work on planaria was first published, there were joking suggestions (I *hope* they were joking) that students eat their professors and get their education that way.

Surely, though, there are alternatives. Perhaps a supply of any kind of RNA would help—just additional raw material. The injection of such "unconditioned" RNA has been reported as having produced borderline improvements in learning ability.

And then, too, why not encourage the body to form greater quantities of RNA for itself? A certain drug called "Cylert" (its chemical name is magnesium pemoline) is known to increase RNA production by 35 to 40 percent. When used on rats, it was found to improve the ease of conditioning markedly.

Experiments of this sort are being conducted (very cautiously) on human beings; specifically, on patients suffering from premature senility. D. Ewen Cameron at Albany Medical Center reports that at least 17 of 24 patients showed improvement.

Upsetting the total euphoria that might follow such positive results is a statement published in the August 5, 1966, issue of "Science" by a number of scientists from eight different laboratories. Their independent attempts to transfer conditioning along with RNA from trained rats to untrained rats have all failed.

This is not, however, grounds for despair or even confusion on the part of those who hope for startling advances. Scientists are now in mid-leap, so to speak, in this field of inquiry, and it is a particularly subtle and complicated field. Different laboratories generally conduct complex experiments with variations that seem unimportant but could turn out to be crucial when all the facts are in. The measurement of learning ability is, in addition, a particularly tricky process and what seems like learning to one experimenter may not seem so to another.

The paper in "Science" puts it this way: "Failure to reproduce results is not, after all, unusual in the early phase of research when all relevant variables are as yet unspecified."

The negative results do not necessarily indicate that RNA is not involved in the mechanism of memory, or even that such memory cannot be transferred. What it does indicate is that the technique of transference has certainly not been perfected as yet; and at this early stage of the game, that is not surprising.

One can't consider RNA molecules by themselves. They come from somewhere. It is known, for instance, that specific RNA molecules are formed as "copies" of similar, but even more complicated molecules, called DNA, in the cell nucleus. Fresh combinations of RNA molecules are not known to be formed within the cell in any other way, and many scientists doubt that incoming sensations can form RNA molecules directly.

The DNA molecules make up the genes, or units of heredity, and these are passed along from parents to offspring by means of a complex but nearly foolproof mechanism.

Each cell contains a long chain of DNA molecules, with each part of those molecules capable of producing an RNA copy of a certain structure. It may be that some of the portions of the DNA molecules are ready to serve as models from the start, and through these portions the cells can form the RNA types they need for the ordinary working of their chemical machinery.

Other portions of the DNA molecules are perhaps blocked off to begin with. A particular sensation might then act to unblock a particular section of one of the DNA molecules, and an RNA molecule matching that unblocked section would then be formed.

This would mean that each person carries a vast supply of possible memories, a "memory bank," so to speak, in the DNA molecules he was born with—a supply vast enough to take care of all reasonable contingencies. The nature of the memory bank might be quite similar among individuals within a species or even within closely related species. It would then become understandable why an RNA molecule producing a particular memory in one will produce a similar memory in the other, and why learning might conceivably be transferable.

And if RNA isn't the beginning, it isn't the end, either. The chief function of the RNA molecules, as far as we know, is to bring about the formation of protein molecules. Each different RNA molecule is involved in the formation of a different protein molecule. Could it be the protein molecule rather than the RNA that is truly related to the memory function?

One way of testing this is to make use of a drug called puromycin. It interferes with the chemical machinery by which the cell produces protein through RNA, but doesn't affect the formation of RNA itself.

Louis B. Flexner and his wife, Josepha, conducted experiments with puromycin at the University of Pennsylvania School of Medicine. First, they would condition mice in a simple maze, teaching them to follow Path A to avoid a shock. The conditioned mice were then given an injection of puromycin and promptly forgot what they had learned. The RNA molecule was still there but the key protein molecule could not be formed. (Once the effects of the puromycin wore off, the mice could be retrained.)

The memory loss depended on when the injection of puromycin was given. If the Flexners waited more than five days, puromycin did not induce forgetting. It was as though something permanent had been formed; as though short-term memory had become long-term and only the former could be affected by puromycin.

Another example involved reversal learning. The mouse, having learned to follow Path A to avoid shock, is suddenly shocked every time it enters Path A. To avoid the shock, the mice had to take Path B. Once the mice had learned the reversal, puromycin was injected. The Path B memory, still short-term, was wiped out, but the Path A memory, long-term by now, was unaffected. The mice returned to the pattern of taking Path A.

All this can be connected with the current-cycles mentioned at the beginning of the article. Suppose that when RNA forms proteins, those proteins contribute to the formation of new dendrites or, perhaps, to the activation of old ones. If this is a gradually strengthening effect, then for the first few days the new current-cycles are feeble and can easily be broken if the flow of specific protein is interfered with, as by puromycin injections. Eventually, though, the dendrites would have been built up to the point where the current-cycle is firm and intense and requires no further protein. Puromycin, after that, is without effect.

But that implies that short-term memory is converted to long-term memory. What if it is the other way around?

Jacobson (who transfers RNA from creature to creature) conditioned planaria and then reverse-conditioned them. He transferred their RNA into new planaria and found that he transferred the tendency to conditioning, but *not* to the reverse-conditioning.

It could be that when a planaria is made to "forget" something it had learned, the RNA molecules formed in the learning process are not removed, merely blocked in some fashion. If the RNA alone is transferred and not the blocking agent (whatever it might be), then only the memory is transferred and not the forgetting. This would back the hypothesis that all memory is long-term and that nerve cells spend their time arranging a forgetting mechanism rather than a remembering one—rather *à la* Freud.

All this work on memory is exciting in the extreme and holds out all sorts of hopes (and fears) for the future. Can we improve our memories by taking pills? Can we learn faster and grow more educated through chemical stimulation? Can we even become more intelligent? Can we adjust minds, by external manipulation, to suit our needs? Can we, through our own effort, change Man into Betterman? Or will some of us decide that what is needed is something else and change Man into Docileman?

But the consequences, whether for good or evil, are perhaps not imminent. With all the excitement of these last few years it may nevertheless be that we are but on the first step of a rocky road that fades off into the farthest visible horizon.

Chapter 3   The Hungry People

It is wonderfully easy to preach to those who are overweight. You can scare them with the possibility of an early death and brusquely order them to eat less. You can tell them kindly to try special exercises, such as pushing the chair away from the table halfway through the meal or turning the head briskly left, right, left when offered a second helping.

Nothing sounds simpler than to follow such advice. Why, then, do so many people go on gaining, even though overweight is uncomfortable, is considered unattractive, and is a danger to health? What makes the plump person keep eating?

An explanation that is popular with a lot of people is that overweight is essentially a matter of psychology. The fact may be visible all over the body but the cause, according to many psychiatrists, is hidden in the unconscious. Overweight is of "psychogenic origin."

If this is true, those moderately fat people who are not actually suffering from some obvious hormone disorder are the victims of personality problems that force them to overeat —against the advice of friends and doctors and against their own common sense and often against their most compelling conscious desires.

Perhaps they were overprotected and overfed as children until the habit of overeating was irrevocably established. Perhaps, on the other hand, they were rejected, and turned to food as compensation. Perhaps the trauma of weaning drove them to seek solace in eating. Or perhaps they found themselves trapped in a period of oral eroticism that they never outgrew. Or—still more complicated—perhaps they eat compulsively to conceal from themselves an even more deeply hidden desire to *reject* food and mother.

Psychiatrists are certainly at no loss for explanations and psychoanalysis would seem to offer the possibility of a cure. Yet the increasing number of psychiatrists in the last gen-

First published in *Mademoiselle*, October 1960.

eration or two cannot be said to have resulted in any noticeable defeat of overweight. On the contrary, more people are overweight than ever. If we are to judge by results, then there would certainly seem to be serious shortcomings to the psychological approach.

Some years ago, two investigators at Iowa State College reported an attempt to check the theory of the psychogenic origin of overweight. They studied over a hundred girls going to rural schools, dividing them into those who had been distinctly fat for at least three years and those of normal weight.

To prove the psychogenic theory correct, the fat girls should have shown more signs of emotional disturbance than the girls of normal weight: their schoolwork should have been below par and they should have done badly in tests designed to assess mental stability, sexual attitudes, and so on.

But when comparisons were made the fat girls showed no difference as a group. Their schoolwork, their stability, their sexual attitudes were indistinguishable from those of the slender girls. In fact, the experimenters could find only one definite difference between the two groups. The parents of the fat girls were, on the average, markedly stouter than the parents of the girls of normal weight.

This last point is not surprising. Earlier studies of a large number of cases have shown that only 10 percent of the children of parents of normal weight grow fat. When one of the parents is fat, 50 percent of the children tend to be overweight. When both parents are fat, 80 percent of their children will share the same fate.

This seems to indicate that overeating may be a result of parental example. And yet identical twins tend to weigh about the same, even when they are brought up separately and exposed to different eating habits.

There is good reason, therefore, to look squarely and somberly at inheritance. There may be an inherited quirk in the body's physical make-up that leads to overeating and the predominating cause of overweight may well be physiological.

Some nutritionists recognize this and complain that the search for the physiological causes has been hampered by the popular attitude toward overweight. Those who are not fat (and this includes some doctors and nutritionists) seem too often to assume that overeating can be avoided by the pure and simple use of will power.

A failure to exert the will in this respect is "gluttony." To find a physical cause for overeating, after all, exonerates the glutton—and this seems almost immoral to some people.

Yet it is impossible to rule out physiology. An inherited tendency to overweight has been recognized and studied in animals, in whom complex psychological motivations are unlikely to be involved. There are strains of laboratory mice that, when allowed to eat freely, do so until they are twice the weight of ordinary mice (who eat more sparingly even when supplied with unlimited food). This "fat tendency" is inherited and can be followed from generation to generation:

Our domestic animals are in many cases deliberately bred in such a way as to develop just those strains that have a pronounced fat tendency. The domestic pig is little more than a fat-making machine and is nothing like the thin, rangy wild pig from which it is descended. Why not consider inherited factors involving overweight in humans too? And why not ask whether they involve some defective working of the body machinery?

We all know that food intake is regulated by the appetite. You eat when you are hungry and stop eating when you are full. For most people these automatic adjustments work well enough to keep body weight steady (within 2 or 3 percent) indefinitely. These lucky ones needn't be particularly conscious of what or of how much they eat. Their weight just takes care of itself.

But this is not true for everyone. There are those who find themselves steadily gaining weight if they pay no attention to their diet. To keep from getting fat they must make a conscious effort to restrict their intake, constantly eating less than they want, sometimes to the point where life is a misery to them.

The person whose appetite exceeds his body's needs may find a plateau of moderate overweight from which he will not vary much. As he gains weight he will have to lift, tug, and move those extra pounds at every step and with every motion. This will mean the expenditure of additional energy and it may be enough to balance his moderate overeating. In other cases, however, an overweight person will eat more to make up for the extra energy so expended and will go on gaining, perhaps slowly, until such time as he deliberately decides to do something about it.

Nor need it be a question of absolute food intake. One recent study of schoolchildren showed that most of the plump ones actually ate less than did those of normal weight. But they were also more inclined to sit still for hours watching television, while the others devoted their leisure to livelier activities.

It is the balance between food intake and energy expenditure that determines weight gain or loss. Among overweight people there is a constant tendency to eat just a bit more food than is necessary to replace the energy expended, however little or much that expenditure may be. That "little bit more" is laid away as fat.

May it not be, then, that with such people there is something wrong with the appetite control? Suppose we compare the appetite control to the thermosat on a furnace (indeed, the appetite control is frequently called the "appestat" by nutritionists). Just as the thermostat may be set for different temperatures—and may therefore be set at one that keeps a room too warm for comfort—so may an appestat be set at different levels. The person whose appestat is set too high is one of the Hungry People. He gets hungry sooner and stays hungry longer, and before long he gets fat.

This is sad, for in America we have come to think of fat as ugly and we *know* that it is a danger to health. Fat people are about four times as likely to develop diabetes as people of normal weight and nearly twice as likely to develop diseases of the heart and circulatory system. For the sake of both health and appearance many fat people try to lose weight, usually by going on a diet. But for fat people whose appestat is set high this can be simply torture. To make matters worse, their bodies automatically compensate for restricted food by restricting activity, so that in spite of the agony they may actually lose less weight than would an ordinary person eating the same amount.

A fat person on a diet is pushing down the appestat

manually, so to speak. He has to have his finger on the control all the time, because as soon as he relaxes it moves back to its automatically set position and he begins to gain again. The woods are full of fat people who have lost weight on strenuous diets and have then gained it all back again.

You can tamper with the appestat by means other than the naked force of sheer will power. You can take pills that deaden the appetite. Or you can try to "fool" the control by eating very slowly or in small nibbles during the day. There are trick diets like those involving high-fat, low-carbohydrate foods, since fat, the theory goes, seems to deaden the appetite faster and for longer periods than carbohydrate does. But however you do it, once you have achieved the weight you want and give up whatever trick you have been using, then —very likely—up you go again.

But where is the appestat to be found and how does it work? It seems to be located in a part of the brain called the hypothalamus (see Chapter 1). If the hypothalamus of a laboratory animal is damaged chemically or surgically the appestat is shoved drastically upward. The animal begins to eat voraciously and soon gets fat.

As to *how* the appestat works, there is considerably more dispute. If its vagaries are controlled not by personality disturbances but by something physical and material, what is it?

One interesting possibility arises from a theory put forward by Jean Mayer, a physiologist at Harvard Medical School. This theory involves the quantity of glucose in the blood. Glucose is a type of sugar always found in the blood in small quantities, and it is stored in the liver as a starchlike substance called glycogen. The body's cells absorb glucose out of the blood stream and use it for energy. As glucose is used up, more is produced in the liver from its stores of glycogen and is dribbled into the blood at just the rate needed to compensate for its withdrawal by the cells. When you are active, the cells are using up larger amounts of glucose and, to make up for this, liver glycogen is converted to glucose at a faster rate. When you are resting, the cells are using less glucose, and glycogen conversion slows up too. The result is a well-controlled balance. But the balance is not perfectly steady. The glucose in the blood slowly decreases a little during a fast and rises again after a meal. Mayer suggests that this variation is what affects the appestat in the hypothalamus. The cells of the appestat are continually testing the glucose level of the blood. As the level dips, the appetite is turned on: as it rises, the appetite is turned off.

If Mayer's theory is accepted as a working hypothesis, the next question is: What regulates the quantity of glucose in the blood and so neatly keeps the balance between the opposing tendencies of glucose formation and glucose absorption?

As far as we know, the balance is controlled chiefly by the activities of two hormones produced by certain cells in the pancreas. One of these is a well-known hormone, insulin. Insulin tends to keep the glucose level low, apparently by making it possible for cells to absorb glucose easily. If for any reason the glucose level threatens to rise too high, more insulin is produced and is poured into the blood stream. The cells absorb glucose more quickly as a consequence and the level falls again.

The second hormone is glucagon and it works in the opposite direction. It tends to keep the glucose level high, apparently by encouraging the conversion into glucose of the glycogen stored in the liver. If the glucose level threatens to fall too low, the pancreas gets to work, produces glucagon, which converts glycogen into glucose, which is poured into the blood stream so that the level is raised again. With both hormones working smoothly the glucose level is kept steady, except for the very minor fluctuations used by the hypothalamus to run the appetite control.

But what if the hormones get out of order?

Fairly often (too often) the body loses its capacity to form insulin in the required amounts. The tendency to suffer this loss is an inherited characteristic and the condition that results is called "diabetes millitus."

With insulin formed in below-normal quantities, the body cells cannot readily absorb glucose; the glucose level in the blood rises dangerously as a result. In spite of this high blood level, Mayer points out, the cells of the appestat can only absorb a small quantity of glucose, since insulin is in low supply. They behave, therefore, as though only a low level of glucose were present and the appetite is turned on. A diabetic, for that reason, is always hungry and if left to his own devices will overeat. (Because he utilizes the glucose derived from food inefficiently, he nevertheless loses weight.)

And what about glucagon, the other hormone involved in maintaining the blood's glucose level? It has been reported that injections of glucagon will cause a rise in the glucose level and a consequent prompt loss of appetite. This fits in with the Mayer theory.

It is tempting at this point to speculate. Could a decline in the body's ability to form glucagon keep the blood level of glucose too low and therefore set the appestat too high? Are those people suffering from ordinary overweight also those with a tendency to low glucagon? Is this tendency inherited?

If so, does the insulin control go out of order more easily when the glucagon control that co-operates with it is *already* out of order? Is that the reason why fat people are so much more likely to become diabetic than are people of normal weight? Is it possible that simple overweight may someday be controlled by some form of hormone therapy, just as diabetes is? To all these questions we can only say that as yet we don't know.

But whatever the answers, it is useless to consider a fat person weak-willed, mentally disturbed, or merely gluttonous. Scoldings, scare psychology, dream analysis, rarely help except temporarily. Even if Mayer's theory proves wrong in detail, it seems almost certain that some physiological mechanism is involved and, one hopes, will eventually be discovered and understood. When it is, a rational hormone therapy may be evolved and overweight can be treated as the physical disease it is. And meanwhile?

There is a way. People who want to do so can lose weight most safely and permanently if they realize that above all they must be patient. They can never relax completely and if they are to keep the appestat "on manual" indefinitely they had better exert a gentle rather than a heavy pressure.

They can undergo the spectacular procedure of a crash diet or a gymnasium treatment, surprise their friends, and delight themselves with the short-term results. But to what end, if the hard pressure on the manual control is released sooner or later (usually sooner) and the body resumes its regimen of overeating?

It is better to eat a *little* less at each meal than impulse would suggest and to do that constantly. Add to this a *little* more exercise or activity than impulse suggests and keep that up constantly too. A few less calories taken in each day and a few more used up will decrease weight, slowly, to be sure, but without undue misery. And with better long-range results too, for if followed faithfully such a gentle pressure on the appestat becomes easier, not harder, to maintain with time.

At least one extended investigation has shown that if a once-overweight person can maintain a normal weight for six months to a year, he is likely to continue doing so for a long time to come. It is as if the pressure of manual control on the appestat, held gently and persistently enough, becomes, with long habit, so easy to maintain that it is tantamount to a new and lower automatic setting.

Take heart, then. Slow *but steady* may win the race after all.

Special Note: At the time I first wrote the chapter above, I was myself some forty pounds overweight and quite cheerful about it. I was in perfect health, with no signs of diabetes or circulatory disorders. Moreover, I was full of energy, usually in good spirits, and, generally, saw no reason why I should limit my pleasure in good food.

Having written the chapter, though, I began to reconsider. Encouraged by those who love me, I put my preaching into practice. Eventually, I found I had lost thirty pounds. I have now kept myself at this new weight-level for over two years without much trouble. I am, of course, on a perpetual diet in the sense that I no longer eat as much as in the old carefree days, and hope never to do so, but it isn't hard.

Of course, there are still the remaining ten pounds. Well, having had to go over this chapter a second time, I will draw a deep breath and try again.

Chapter 4   Blood Will Tell

We carry in our veins a personal encyclopedia which doctors and biochemists are gradually learning to read. They are still struggling with the obscurer passages, but what they've now deciphered has already prolonged human life.

"Blood will tell" runs the old saw, with a meaning that is all wrong. For what blood was supposed to tell in the old days involved such things as manners and breeding, bravery and honesty—or the reverse; in short, all the things that are determined by training and environment and not by heredity at all.

It was only with the dawn of the 20th century that the true code of the blood began to yield to investigation. Blood will tell, indeed, it was found, *if* it is asked the right questions.

It was discovered in 1901 that blood existed in four major types. These types gave no outward sign. No one could tell, by looking at another person, or by studying any part of him but his blood, which blood type he belonged to.

The difference showed up thus: When different types of blood are mixed, the red blood cells of one of the samples clump together into a sticky mass. This never happens when two samples of blood of the same type are mixed.

In the test tube such clumping is interesting. In the living veins of a human being, however, it can be fatal, because clumps of red blood cells plug up vital capillaries in the kidney, heart, or brain.

Here, at last, was the answer to the erratic results that had accompanied pre-20th century attempts at blood transfusion. Every once in a while throughout history, some doctor had tried to make up for blood loss by feeding blood from some donor into the veins of a patient. Sometimes the patient was helped; often he was killed.

In the 20th century, transfusion became safe and routine. It was only necessary to use a donor of the same blood

This appeared in *Think,* April 1962.

type as the patient or, at worst, of another blood type known to be compatible with the patient's.

The blood types are inherited according to a fixed pattern, so that blood will tell not only the transfusion possibilities but also something about relationships.

Thus, a man and wife, both of blood type A, cannot possibly have a child of blood type B. If a child of theirs does have a blood type B, then only two possibilities exist: either the child was inadvertently switched in the hospital, or the husband in the case is not the real father. It doesn't matter how feverishly the various relatives may testify that the child has his mother's husband's nose and chin. That may be so, but it would be pure coincidence. Blood will tell in such a case, and it doesn't lie. (Though one must admit that it is possible for technicians to make a mistake in determining blood type.)

Relationships of a broader nature are also hidden in the blood. For instance, a blood type commonly called "Rh negative" occurs in an appreciable number of Europeans and among their descendants in other continents. It hardly ever appears in the natives of Asia, Africa, Australia, and the Americas.

In Europe, the incidence of this blood type is highest among the Basques of the Spanish Pyrenees, where one-third of the population is Rh negative. It is possible, then, that the Basques represent an older stratum of European population, submerged by a flood of later immigrants from Asia or North Africa who have been the "Europeans" now for many thousands of years.

That the Basques are the last remnant of the "Old Europeans" is indicated by the fact that their language is not related to any other language on earth—and by their blood.

In fact, the changing ratio of the different blood groups as one travels the face of the globe has been used to try to follow past migrations. A tide of blood type B slowly recedes as one travels westward through Europe from the Urals toward the Atlantic, and this may mark the passage of such Asian invaders as the Huns and Mongols, for the percentage of B is highest in central Asia. Invasions of Australia from the north, of Japan from the west, can also be traced in the blood of the population.

If only the chief blood types, those of importance in transfusion, are used for such purposes, however, we are confining ourselves to a fuzzy technique. These types are spread too broadly and the ratios of one to another differ by too narrow a margin.

Fortunately, many additional blood types have been discovered in the last half century. None of the additional ones are of importance to transfusion, to be sure, but all are clearly definable and all are inherited according to some fixed pattern.

As of now, in fact, over 60 blood types of one sort or another have been identified. The number of possible combinations of those types that might exist in a particular human being (even allowing for the fact that some types are uncommon) has been estimated at 1,152,900,000,000,000,000.

This huge number is 400 million times as large as the total population of the earth. It is therefore quite likely that a laboratory that was equipped to test for all possible blood types (and none is so equipped at the moment, unfortunately) could differentiate any human being's blood from that of any other—except in the case of identical twins.

Potentially, then, you carry your calling card about with you at all times. Once your blood type pattern is recorded in full, it is a case of "John Doe, I presume" to any blood technician who carries through the necessary tests.

And, as a result of a full analysis, relationships, in the narrow sense of paternity, or in the broader sense of tribal migrations, might be worked out with precision. Such problems as the wanderings of the Polynesisns or the route of entry of the Indians into the Americas might be solved once and for all.

(Anthropologists interested in such past wanderings, however, had better work things out in a hurry. The automobile hastened the mixing of peoples within nations, and now the jet plane is introducing a new ease of shifting and uprooting on an intercontinental scale. If this continues, migratory history may be blurred out of all recognition in a few generations.)

All this, in essence, means that blood can, at least potentially, tell us who we are.

It would be interesting if, in addition, it could tell us

what we are. Suppose it could tell us if we are well or ill, for instance; and if ill, how ill and in what fashion. It would be even more fascinating if it could predict the future, and tell us if we are likely to become ill; and if so, in what way.

All this, blood can indeed do, at least potentially. It will give the answers if the proper questions are asked.

Nor is this just a matter of curiosity; just mankind's primitive urge to look into a crystal ball. We are all a prey to sickness, and the more we know about it the more likely we are to arrest its progress, or even reverse it. Best of all, we may prevent it in the first place.

It is a general rule that the earlier in its progress any disease is detected, the easier it is to treat. Now any disease, if it progresses far enough, will produce visible symptoms—or it would not be recognized as a disease. But the body struggles hard to retain its equilibrium against the onslaught of the disease, and by the time symptoms are actually visible, the body has lost the battle, at least temporarily. For best treatment, therefore, the disease ought to be discovered well before symptoms are visible to the naked eye.

Well, every phase of the body's activities, both in health and sickness, is reflected in the complex chemistry of the blood. So we turn to the blood. Diabetes is a good example here.

The advanced diabetic loses weight though he eats voraciously (see Chapter 3); he must both drink and urinate copiously. He is plagued by boils and itching and a number of other more serious, if less immediately noticeable, disorders. By that time, the diabetic is in a bad way and almost past help.

Diabetes is caused by a shortage of a hormone: insulin. This controls the concentration level in the blood of a form of sugar called glucose. As the supply of insulin declines, glucose concentration rises until some spills over into the urine. Detecting the first signs of glucose in the urine will prove the existence of diabetes before the patient is at the last gasp.

But that is still too late in the course of the disease for comfort. One can test the blood directly and note whether the glucose concentration, though not high enough to spill over into the urine, is yet above normal. Better still, one can put the body's chemical devices for handling glucose under a strain. Then we can observe whether the body, while still able to control glucose level under ordinary conditions, will show signs of failure in an emergency. If the body does show such signs, we will be detecting diabetes at the very start.

This is done by means of a "glucose tolerance test." The patient is given a fairly large dose of glucose solution to drink, and his blood is analyzed before and at various times afterward.

The glucose is absorbed quickly through the intestines and floods into the blood. The glucose concentration shoots up as a result. In response to that rise in glucose concentration, however, insulin is produced in higher quantities than normal and the glucose level is brought down to the proper place in fairly short order. In normal individuals, the glucose concentration is about 100 milligrams per 100 milliliters of blood. About 45 minutes after a glucose meal, the figure has risen to 200, but within an hour after that it is back to 100.

If the figure rises markedly higher than 200 after the meal and takes a number of hours to straggle back to normal, it means that the body is having difficulty producing insulin in emergency quantities and there is therefore a good chance that diabetes is developing. When the disease is caught at this stage, a judicious diet and a proper regimen of exercise can maintain reasonable normality indefinitely. The use of insulin injections can well be avoided.

A second example involves the thyroid gland, which controls the rate at which the chemical machinery of the body is grinding away. This is called the "basal metabolic rate," or BMR. Until a few years ago, the way of measuring the BMR was to have a patient breathe from an oxygen cylinder; since the rate at which oxygen was consumed was the rough measure of this BMR. But the thyroid gland produces certain hormones that control the BMR. These hormones contain iodine atoms that are transported by the proteins of the blood stream from the thyroid gland to the rest of the body. Once a method for determining the "protein-bound iodine" (PBI) of the blood was worked out, the slow oxygen-breathing test became obsolete. A quick puncture, and a little blood will tell.

Kidney disease, like diabetes, is easy to detect when it is well advanced. Something is needed to detect it at the start. Well, the prime function of the kidneys is to filter waste out of the blood, and the most important waste is one called urea. It is not difficult to measure the concentration of urea in the blood, and when that goes up beyond the normal stage, the kidney may be starting to fall down on its job—in time to be caught, perhaps.

The liver is the busiest chemical factory in the body and its proper functioning is vital to life. But every necessary substance it forms must be distributed by way of the blood stream, and from the rise or fall in concentration of these, the exact story of the liver's shortcomings, if any, can be worked out. Jaundice is a condition, for instance, in which a yellow-green pigment called bilirubin occurs in the blood in abnormally high concentration. It can be caused by trouble with the red blood cells which, in breaking down too quickly, form abnormal amounts of bilirubin. Or it can be caused by trouble with the liver, which may be blocked from delivering bilirubin into the intestines, as it should, and may be delivering it into the blood stream instead. By checking for bilirubin by two different chemical methods, the biochemist can at once distinguish whether the trouble is in the blood or in the liver.

If the blood is an open book, it is a rather complicated one. Biochemists can detect any of the many dozens of substances in the blood, and a variation in concentration of any one of these may possibly be symptomatic of any of a number of diseases. A rise in a protein called serum amylase may bespeak pancreatitis; a rise in another called alkaline phosphatase may be pointing the finger at bone cancer; a rise in acid phosphatase is possibly symptomatic of cancer of the prostate. A protein called transaminase may, when elevated in concentration, indicate heart damage. A rise in certain types of fatty substances make atherosclerosis an uneasy possibility. There are dozens of other examples.

No one test is a certain indication of one disease, but each test narrows the field of possibilities, and a combination of tests may narrow it quite a bit. The physician is told quite clearly in which direction to look, and he is told this at a time when no open symptoms may yet have had time to develop and when treatment still offers high hopes of a cure or, at least, a stay in the progress of the disease.

What does the future hold? There is every reason to think that blood's value as a diagnostic tool will continue to rise. Since World War II, newer techniques have been continuously devised for analyzing more and more complex mixtures with greater and greater precision. We can take blood apart into its components with more and more certainty.

But not all variations in blood composition are necessarily pathological. The blood groups are a good example of this. As far as we know, a person of blood group A is as normal as one with blood group B, and as likely to live as long and as healthy a life. But the two individuals are different just the same, and when transfusion is involved this difference must be taken into account.

There may be other differences, too, lying within the limits of normality and yet requiring slight gradations of treatment.

For instance, one of the most important functions of the blood is that of supplying to the various cells the substances required for building tissue. The prime requirements here are some twenty closely related compounds called "amino acids." These may occur separately, or combined into giant molecules—the proteins. The particular amino acid make-up of the blood of a particular individual may have important significance to his medical treatment (see Chapter 5).

So perhaps the next century will see human biochemistry become a truly individual matter. A person's blood will not merely be his or her calling card. It will become a record of his past, present, and future history.

The Sherlock Holmes of the future will be the blood technician. Indeed, one can imagine a time when blood analysis will be so perfected through the use of micro-tests, analyzed by computer perhaps, that a drop of blood will suffice to tell your fortune, like the card in the weighing machine. Only it won't tell you that you are to meet an interesting stranger, or are about to go on a long trip. Instead, it will advise you on your diet, tell you of the dangers that may befall, of the little maladjustments of your body machinery which, if ignored, may become serious maladjustments. To your grandchildren, information from that drop of blood may be the key to a healthy and prolonged life.

Chapter 5 The Chemical You

We take it for granted that no two people look exactly alike. A child has no difficulty recognizing its mother, and a boy assures his sweetheart that no one in the world is anything like her. Even identical twins have their differences. And what is apparent to our own sense of sight is equally apparent to the dog's sense of smell. But looks are only skin-deep, our poets tell us. And odor, too, say our television commercials.

Can we go deeper, then? Are there differences in the inner workings of the body that can make themselves apparent in the coldly impartial world of the chemist's test tube? To be sure, all of us use hemoglobin to absorb oxygen and certain enzymes to produce energy. We all have lungs and heart and kidneys. We can all live on the same food, suffer the same diseases, and come equally to the death in the end. But there is more to it than that.

In the previous chapter, we discussed the blood's role in displaying man's chemical individuality—now let's go farther.

In the first couple of decades of this century, an English physician, Archibald E. Garrod, studied the pattern of metabolism in human beings. That is, he studied the sequence of chemical reactions by which the body broke down food to form energy and build tissue. He found cases of people who lacked the ability to bring about one particular reaction or another, with results that were occasionally disastrous (see page 21 for an example).

Such chemical quirks are with a person from birth. The equipment, or lack of it, with which you must conduct your internal chemistry is yours from the start (at least

First published in *Mademoiselle*, January 1963.

potentially, for in some cases the deficiency makes itself fully apparent only in later life). Garrod referred to deviations from what seemed normal metabolism as "inborn errors of metabolism."

Naturally, the inborn errors that are easiest to see are just the ones that produce serious diseases, such as diabetes (see Chapter 4); or startling symptoms, like the relatively harmless alkaptonuria, in which the urine, under certain conditions, turns black.

Realizing that the chemical mechanisms within cells are highly complex, Garrod felt that there might be any number of deviations from the normal that produced neither startling nor dangerous symptoms. In other words each individual might follow a chemical path not quite like that of anyone else and be very little the worse for it. From that point of view we would all be individuals not only in appearance, but in chemistry.

Consider! The body builds up special defensive proteins (antibodies) that react with foreign molecules and neutralize them. This is one of its best defenses against invading bacteria and viruses. Once someone has formed an antibody against the measles virus, he is immune to further attacks of measles. The Sabin vaccine encourages the body to form antibodies against the polio virus by presenting it with the necessary virus in a form that will not actually produce the disease. The body is then made immune without having to undergo the prior risk of polio itself.

A negative instance of the same use of proteins is the fact that the body may also accidentally become sensitive to foreign substances that are fairly harmless in themselves; to the proteins of certain types of pollen, for instance, or to certain types of food. A person will in such a case suffer from hay fever or food allergy.

A particular antibody can distinguish between one foreign substance and another (between chicken protein and duck protein, for instance) even when the difference is not readily apparent to the chemist. It can always distinguish between a foreign substance and the molecules present in the body it belongs to.

If an antibody can distinguish between two proteins, those two proteins must in some way be different. That being so, no two human beings, except for identical twins, can have proteins that are completely alike. The proof of this is that a skin graft will fail unless it is taken from another part of the patient's own body, or, at furthest remove, from the body of his identical twin, if he is lucky enough to have one.

The patient's body will recognize and form antibodies against the skin protein of any human being other than himself (or his identical twin). Those antibodies will prevent the graft from "taking" and will, at great inconvenience or danger to himself, prove the patient a chemical individual.

So far, much of our medical magic is confined to methods that strike at the common denominator of all mankind. Aspirin will relieve pain in almost anyone, and penicillin will almost always halt the growth of certain germs in anyone. A physician must, of course, be on the lookout for the tiny minority who are sensitive to these universal panaceas, but he can, in general, prescribe such drugs freely.

As knowledge increases, however, a finer control, carefully geared to the needs of each individual, will be added to such wholesale therapy. The physician will have to recognize that there is not only a psychological and biological you, but a chemical you as well.

The first step in the direction of finer control will undoubtedly involve proteins. After all, most of the substances that induce an antibody reaction are proteins, and the antibodies themselves are proteins. Obviously, then, proteins present in the body differ among themselves in subtle ways, and the body can design other proteins to take advantage of these differences.

What are these subtle differences? To begin with, proteins consist of large molecules. Even a protein molecule of only average size is made up of a conglomeration of perhaps four hundred thousand atoms. In comparison, a water molecule is made up of three atoms and a molecule of table sugar of forty-five atoms.

Atoms within the protein molecule are arranged in combinations called amino acids, each of which is made up of anywhere from ten to thirty atoms. The amino acids are strung together, like beads in a necklace, to form a protein molecule.

Although the general structure of the amino acids is alike, there are differences in detail. An individual protein will be made of anywhere from fifteen to twenty-two different types of amino acids, placed in a particular order in a chain that may contain up to thousands of them altogether.

Naturally, if two proteins are made up of different numbers of amino acids, they are different, and this difference can be distinguished by antibodies. Again, if they are made up of the same number of amino acids but of different proportions of the various types, they are different.

The interesting thing is, though, that if two protein molecules are made up of the same number of the same types of amino acids, they will *still* be different if the order in which those amino acids occur in the chain differs. It is as if you were to make a necklace out of twenty beads—five red, five yellow, five blue, and five green. Depending on the order in which you arranged them you could make twelve billion different patterns.

But proteins are almost never as simple as that. An average-sized protein would have five hundred amino acids, not twenty; and the amino acids would be of twenty different types, not four. The number of possible ways in which the amino acids of an average protein molecule could be arranged requires well over six hundred zeros to be written.

This being so, it is obvious that everyone in the world can easily have his own individual proteins not quite like those of anyone else. In fact, every living creature who has ever lived can have had his own individual proteins not quite like those of any other creature who ever lived—and, therefore, his own individual chemistry as well.

But then, if every creature has different proteins and the body is sensitive to foreign proteins, how is it that we can eat? Fortunately the food doesn't enter our body in its original form. It remains in the alimentary canal until it undergoes certain digestive changes. Only then does it cross the walls of the intestines and enter the body itself.

Proteins, in being digested, are broken down to individual amino acids, and only the amino acids are absorbed. If even a small bit of the intact protein itself were absorbed, the body would become sensitized to it and would thereafter display a strong allergic reaction to that protein. The amino acids, however, are harmless.

Out of the absorbed amino acids, the body builds the individual proteins of its own tissues, tearing down any amino-acid excess for energy. Naturally it makes use of amino acids in certain proportions for building body protein, and these proportions may not be quite those in which the amino acids occur in the particular food proteins that have been ingested. Fortunately the body can deal with the discrepancy by changing the molecular structure of some amino acids to that of others. One present in excess may be converted to one that is deficient, making for a more efficient mixture.

However, there are limits to this. Almost a hundred years ago, it was discovered that rats would die if their only source of protein was corn, but that they would live if a bit of milk protein were added to the diet. The explanation for this turned out to be that corn protein was deficient in an amino acid called tryptophan, which milk proteins possessed in ample quantity. Apparently the rat could not make tryptophan out of other amino acids, and, being unable to maintain the protein level of its tissues without it, it died.

In the 1930s the American biochemist, William C. Rose, through dietary experiments on university students, found that the human body could not make eight of the amino acids. These eight are called the "essential amino acids," because their presence in the diet is essential to health.

On a reasonably varied diet that is above the starvation level, it is unlikely that many of us will suffer seriously from deficiency of one or more of the essential amino acids. But some may suffer borderline discomforts and malaise because the diet we choose for ourselves supplies us with the wrong pattern of the essential amino acids.

Nutritionists are easily able to analyze foods for their amino-acid content; they also know average daily requirements. It is possible to make blanket decisions about what amino acids are missing from a given diet and how to supplement the deficiency with food or pills. This again is wholesale therapy. But more individual treatment is becoming possible.

There is a fairly simple method of sizing up a patient's

chemical individuality. The body supplies protein for its own cells, which circulates in the blood stream and presumably has the particular pattern of amino acids required by the individual involved. The proteins in a few drops of blood can be analyzed, and the amino acid pattern—a kind of chemical fingerprint—taken. When this is compared with the amino-acid pattern of the patient's diet, specific suggestions for supplementation can be made.

A person with inborn quirks in his pattern can be treated individually and need not be victimized as a member of a mythical normal group. This can be applied on a larger scale to whole areas where a low standard of living requires efficient supplementation.

As medical and chemical knowledge increases, the day may come when each individual is metabolically assayed in infancy and periodically thereafter; when central medical files may contain records of each individual pattern. And then perhaps no doctor will treat any patient, except in emergency, without preliminary study of the pattern.

It isn't humanity in the abstract, after all, that the doctor faces when you go to him for help. It is you, the individual; the chemical you.

Chapter 6   Survival of the Molecular Fittest

As explained in the previous chapter, proteins are one of the chief keys to chemical individuality, thanks to the intricacies of their molecular structure. And among the most significant proteins are the various enzymes, mention of which was made in Chapter 1.

It is no wonder, then, that biochemists are eagerly attacking the problem of enzyme structure, and achieving results, too.

Enzymes, like all proteins, are built up of relatively simple units—amino acids. The twenty or so different types of amino acids occur in the smaller enzyme molecules to the

First published under the title "The New Enzymology" in *Consultant*, May 1965. Copyright © 1965, reprinted by permission of Smith, Kline & French Laboratories.

extent of two to six each; in the larger ones to the extent of several dozen each.

Biochemists know the detailed structure of each different amino acid. They also know the exact manner in which one amino acid is connected to another to build up a "peptide chain." To work out the exact formula of a protein, we must first determine which amino acids, and how many of each, are in its particular peptide chain.

The peptide chain may be hydrolyzed by heating it in an acid solution. This process decomposes the chain into individual amino acids. The mixture can then be analyzed, and the number of each variety of amino acids present in the chain determined.

This, however, is not enough. In what order are the amino acids present in the peptide chain? The number of possible orders is very great in even the simplest protein. For example, there is a hormone called oxytocin which is one of the smallest of the naturally occurring proteins. Its molecule is made up of a peptide chain containing only one each of merely eight different kinds of amino acids. Yet these eight amino acids can be arranged in no less than 80,220 ways.

The situation grows inconceivably complex for the large protein molecules—and yet not hopeless. By chopping peptide chains into small pieces, containing two or three amino acids each, and working out the order in one small piece at a time, the order in the entire chain can be deduced. By 1953, the order of the fifty-odd amino acids in the molecules of the hormone insulin (see Chapter 4) had been worked out completely.

Insulin was the first protein molecule to be conquered in this fashion, and the conquest had taken eight years. With the detailed technique worked out, however, larger molecules were conquered in less time. Thus, the molecule of ribonuclease (an enzyme which brings about the breakup of ribonucleic acid—the famous RNA concerning which there was so much said in Chapter 2), made up of a peptide chain of 124 amino acids, was soon completely worked out.

The manner in which ribonuclease (or any enzyme) brings about a chemical reaction is itself quite subtle and interesting, and I will have more to say about that in the next chapter. Naturally, once chemists knew the exact amino acid make-up of an enzyme molecule, they were interested in knowing what there was about that make-up that gave the molecule such amazing facility at bringing about a certain chemical reaction, such as, in this case, the breakup of the RNA molecule.

Carefully, chemists went on, therefore, to alter this or that particular amino acid in ribonuclease in order to find the "active sites," the portions that were directly involved in the enzyme action. It turned out that some particular amino acids, when altered ever so slightly, were associated with a loss in enzyme activity, while others could be manhandled quite a bit without any overall effect. The key spots turned out to be amino acid 12 (of the variety called "histidine"), amino acid 41 ("lysine"), and 119 ("histidine" again).

It seems quite likely that, despite the wide separation of these three amino acids in the chain, this represents a single active site. The peptide chain is not a long, straight rod, after all, but is more like a flexible rope which can be folded in such a way in the ribonuclease molecule as to bring positions 14, 21, and 119 together. In this way a specific three-amino-acid pattern is formed.

The enzyme molecule is held in its folded shape by particular links between particular atom groupings. One of the most important of these involves the amino acid "cystine." Cystine is a sort of double molecule. Each half is a complete amino acid in itself, the two halves being connected by a chain that includes two sulfur atoms (a "disulfide bridge"). One half of the cystine can form part of one peptide chain and the other half part of another. In this way, two separate chains (or two sections of a single chain) are firmly held together by a disulfide bridge.

The ribonuclease molecule has four such bridges linking different parts of the chain. There are other weaker types of attachment also, all contributing to making the peptide chain fold properly, so as to create an active site.

But if a small group of amino acids forms the active site, why the need for over a hundred other amino acids? Some of the reasons for this have already become clear.

If ribonuclease is split into two parts at the position of

amino acid 20, each part, separately, is inactive. If solutions of the two parts are mixed, much of the activity is restored. It is as though the two parts were able to line up properly even though there are countless trillions of possible ways of lining up *improperly*. Apparently, the amino acid arrangement in an enzyme is such that natural folds appear in the chain, folds that bring the proper amino acids together and form an appropriate active site. It would seem, then, that the long chain is necessary in order to devise a natural folding process leading to the automatic formation of that active site.

But why build a long chain merely to make the active site a sure thing? Why not have the amino acids of the active site put together permanently and dispense with the rest of the molecule altogether? For one thing, it is not desirable to have the enzyme active at all times.

Consider the common enzymes trypsin and chymotrypsin. They are digestive enzymes which act upon the food in our intestines, breaking down the protein molecules in the food and converting them to small fragments, which are then broken down to individual amino acids and absorbed.

Such enzymes are part of a complex team and should perform their work only at an appropriate moment. They are therefore secreted in inactive forms called trypsinogen and chymotrypsinogen. The peptide chains in these inactive forms cannot easily fold in such a way as to bring an active site into being. If, however, the chain is broken at one specific point, what is left folds properly and an active enzyme comes into being; chymotrypsinogen becomes chymotrypsin while trypsinogen becomes trypsin.

Similarly, ribonuclease, which must fold to form an active site, will presumably only fold in the proper manner when certain conditions are fulfilled. It can then be inactive or active in order to suit circumstances. An active site, ready-made, would be *always* active, and that would not suit the needs of living tissue, which requires enormously subtle flexibility in the behavior of its components.

Let us return to the digestive enzymes we have just mentioned. The molecule of trypsin contains 223 amino acids divided into three peptide chains held together by cystine bridges. That of chymotrypsin is somewhat larger. The amino-acid order in both enzymes has been worked out.

Trypsin and chymotrypsin turn out to have identical active sites, and about half the amino acid order in trypsin is identical with that in chymotrypsin. In view of this similarity, it is not surprising that the two enzymes are similar in function; that both bring about the breakdown of protein molecules as part of the digestive process.

But there are differences, too. The differences in the amino-acid order make it possible for trypsin to attach itself to protein molecules in one fashion while chymotrypsin attaches itself to them in another fasion. In this way, proteins are oriented differently with respect to the active site and the two enzymes are not *precise* duplicates.

Because of this difference in orientation, trypsin will split only certain types of amino-acid links including those that involve lysine, which I have already mentioned, or another amino acid, arginine, which is rather similar to lysine in certain respects. Chymotrypsin will split links involving amino acids such as phenylalanine, tyrosine, and tryptophan (all three of which have in common the presence of a ring of six carbon atoms in the molecule).

Since both trypsin and chymotrypsin have active sites of identical structure, the purpose of the remainder of the molecule is shown in still another light. By governing the manner in which the enzyme combines with those molecules it influences, additional flexibility of behavior is made possible that would not exist if the active site were present in isolation.

The similarity of trypsin and chymotrypsin suggests that both developed from the same molecular ancestor. The differences between the two enzymes arise from the fact that though the ability to form specific peptide chains is inherited, this ability is occasionally distorted in transit ("mutation").

The process of evolution by natural selection applies to peptide chains, presumably, as well as to organisms as a whole. If a new peptide chain is formed which has an inefficient method of functioning, or none at all, organisms possessing it may tend to die out. A new peptide chain with a slightly altered function, or even a radically new one that

65

can be put to use, will survive, and the organism possessing it may be modified to suit the new function. There is thus a survival-of-the-fittest among molecules as well as among organisms.

What's more, evolution among molecules may shed light upon evolution among organisms. The structure of the molecules of the enzyme cytochrome C (one that is involved with the handling of oxygen in tissues) has recently been studied in 13 different species from man to yeast. About half of the 104 to 108 amino acids of this enzyme were found to be present in identical order in all the species. This is strong evidence in favor of the belief that all of life originates from a common ancestor.

The differences that do exist become more marked as the species are more distant. The peptide chain of the cytochrome C molecule in man differs by only one amino acid from that in a rhesus monkey. There are, however, 21 differences between man and a tuna fish; and 48 differences between man and a yeast cell, as far as this molecule is concerned (even though it performs an analogous function in all the species).

Undoubtedly, if chemists could simplify their techniques to the point where numerous enzymes could be studied in many species, the overall differences would be complex enough to reveal the evolutionary pattern in great detail.

Chapter 7 Enzymes and Metaphor

A classic experiment that often serves to start a course in high school general chemistry is one in which oxygen is prepared by the decomposition of potassium chlorate (which contains oxygen atoms in its molecule). The directions for conducting the experiment are explicit. The student does not simply heat potassium chlorate. Manganese dioxide (which also contains oxygen atoms in its molecule) must be added

This was presented as a paper to the American Chemical Society on April 7, 1959, and was then published in the *Journal of Chemical Education*, November 1959. Copyright © 1959 by Division of Chemical Education, American Chemical Society.

first. Without it, potassium chlorate must be heated strongly and oxygen evolution is, nevertheless, slow. With it, the mixture may be heated gently and oxygen is given off quickly.

It is necessary to explain to the student that the manganese dioxide does not enter into the reaction, if only to prevent him from suspecting that the oxygen comes from the manganese dioxide and that metallic manganese is left behind. The function of manganese dioxide is only to accelerate the breakdown of potassium chlorate in some fashion that does not consume the manganese dioxide. Its mere presence is sufficient. It is a catalyst, and the process of influencing by mere presence is called catalysis.

This, left to itself (as it often is), can lead the beginner to a lifelong and unnecessary association of catalysis with mystery. The notion of influence by mere presence rather than by participation is uncomfortably like a kind of molecular *psi* force, an extrasensory perception on the part of potassium chlorate that the influential aura of manganese dioxide is present, or perhaps a telekinesis, a supernatural action at a distance on the part of the aloof but godlike manganese dioxide molecule.

Any unnecessary aura of mystery in science is undesirable, since science is devoted to making the universe less mysterious, not more so. The fact that any student going on to industrial chemistry will be constantly involved with catalysis and that any student going on to biochemistry will meet with those exceptionally useful protein catalysts, the enzymes, makes this particular unnecessary mystery especially undesirable.

It is, naturally, impossible to stop an introductory course long enough to delve into surface chemistry with the necessary detail to remove the mystery. For one thing, the students lack the necessary background for it, and for another, nothing like this is required. All that need be done at the start is to puncture the mystery; time enough later to supply the rationale.

To do away with mystery, it is only necessary to offer the students common examples of how a reaction can be hastened by mere presence of an extraneous substance; examples that, on the face of it, do not involve witchcraft. In short, a student may be unready for advanced chemistry, but he is always ready for metaphor.

Once given the metaphor, the student will retain it indefinitely, if it is dramatic enough. Even if he never proceeds beyond the elementary chemistry course, he will yet avoid in this one tiny respect the plague of mysticism; and a contribution will have been made to the rational view of the universe, which is one of the ends of scientific thought. If the student proceeds into advanced chemistry courses where catalysis will crop up again and be placed on a firmer theoretical base, he will at least have his proper start and be able to approach the subject with greater confidence.

For instance, how can a catalyst influence a reaction by its mere presence? What is there in ordinary life that can offer an analogy to such an esoteric phenomenon? Suppose we make use of the "brick-and-inclined-plane" metaphor.

Instead of potassium chlorate breaking down and down and down, liberating oxygen, imagine a brick sliding down and down and down a gentle incline, liberating energy. Both are spontaneous processes, but both need initial pushes. The potassium chlorate requires the encouragement of heat; the brick will require an initial thrust by hand.

Suppose the incline on which the brick rests is a rough one, however, so that there is a great deal of friction between brick and surface. Despite gravitational pull and the helping hand, the brick stops quickly once the hand is removed.

Now suppose you were to coat both brick and incline with ice. The brick suddenly slides more easily. A gentle push to start it, or even, perhaps, no push at all, is necessary.

But the ice itself does not push the brick; it does not increase the gravitational force; it does not supply force in any form; it plays no active role at all. Its mere presence is enough. Nor need much be present; only enough to coat thinly those portions of brick and incline that come into contact. Nor is any of the ice used up in the process, ideally. When one brick has moved on, the ice is all there; another iced brick can be set sliding, and then another and another.

A catalyst is defined as a substance capable of accelerating a chemical reaction by its presence in small quantities, without itself undergoing permanent change in the process. Eliminate the word "chemical" and the ice that coats an incline is a perfect catalyst.

An alternative analogy is the "writing-board" metaphor. Imagine a man with a pencil and paper, and nothing else, standing in the midst of a desert with only soft, shifting sand underfoot. The man wishes to write a note upon the paper.

The man knows how to write, he has the wherewithal to write with, and the wherewithal to write upon. Nevertheless he can write only the most fumbling note, one that is very likely undecipherable, and he will almost certainly tear the paper in the process.

Now imagine him suddenly endowed with a smooth writing board of polished wood which will not itself take a pencil mark. How different the situation suddenly.

The man undergoes no increase in his knowledge of writing. His instrument of writing remains as before, the pencil. The only object upon which he can write a note remains the paper.

Yet now his message can be written smoothly, clearly, and painlessly—all thanks to a writing board, which hastens the process by its mere presence and is unchanged in the process. Both paper and pencil are somewhat used up and the man himself has expended some calories, but the writing board has suffered no significant loss. It can be used for an indefinite number of similar jobs. It is, in short, a catalyst.

Both metaphors serve, furthermore, to introduce the notion that catalysis is essentially a surface phenomenon; that a reaction is hastened (whether it is the sliding of a brick, the writing of a note, or, by extension, the breakdown of potassium chlorate) by the provision of a surface that is specifically suited for the activities involved in the reaction.

Later in the course, the student may be presented with the notion that catalysis hastens a reaction without, however, changing the position of the equilibrium point. Suppose, for instance, one begins with two substances A and B, which react to form C and D. Left to themselves the reaction will proceed partway, reaching a halt at an equilibrium point where A, B, C, and D are all present in fixed proportions. The presence of a catalyst hastens the rate at which that equilibrium point is reached, but does not alter its position.

Moreover, if you begin with C and D, they will react partway, reaching an equilibrium point—the *same* equilibrium point—with A, B, C, and D present in fixed proportions. And the same catalyst will hasten this reverse reaction, too.

To the student who first encounters this fact, there may seem something diabolical about an inanimate substance that can pull in either direction as though it knew in advance where the equilibrium point was.

Yet it is simple to demonstrate that the catalyst is not pulling in two directions but in only one, if we go back to our brick-and-inclined-plane metaphor. Imagine a double incline in the shape of a shallow, blunted V, made of a rough, high-friction substance. Again, a coating of ice will serve as catalyst, allowing the brick to slide. Observe though that it will slide down either slope of the V and in both cases end at the same point, the bottom.

If the top of one arm of the V is labeled "A and B" and the top of the other "C and D," and the whole viewed from directly above, it would seem that the catalyst works in either direction. The brick will seem to slide from right to left, or from left to right, and, in each case, will seem to stop at a mysterious midpoint that seems no different from any other point.

View it from the side, though, and you see at once that the reaction is in one direction only—downward, in the direction of the pull of gravity. The ice-catalyst accelerates this one downward movement. The endpoint (or equilibrium) is the lowest point in the V, the point of least gravitational potential, and hence unique. Even the least intuitive student can see that the ice is actuated by no mysterious foreknowledge of the equilibrium position. The brick simply slides to the bottom.

The student should see at once, too, why catalyzing a reversible reaction does not change the equilibrium point and why hastening a reaction in a particular direction does not cause it to move further in that direction. Obviously coating brick and incline with ice to enable the brick to slide more quickly doesn't alter the position of the bottom of the V or allow the brick to go past it and climb halfway up the other side and remain there permanently.

In more advanced courses, the student will learn there is something called "chemical potential" that can be dealt with, in some ways, analogously to the familiar and every-day gravitational potential, and the picture he grasps at the beginning will stand him in good stead later.

Nor need the usefulness of metaphor be restricted to only the more elementary notions. Eventually, it will be explained that a catalyst achieves its results by lowering the energy of activation.

In other words, the substance being worked on by the enzyme first forms an unstable intermediate compound which then breaks down to form the final product. The unstable intermediate compound requires the input of a comparatively large amount of energy, but until it is formed, no final prod-ucts will be formed, although the products themselves are not particularly high-energy. The entire reaction will pro-ceed no more quickly than the unstable intermediate can be formed.

The catalyst, by making the intermediate more stable, allows its formation with a smaller energy input. This hastens the rate of formation of intermediate and consequently has-tens the reaction as a whole.

Often the energy of activation (the energy required to form the intermediate) is represented as an "energy hump" between products and reactants. The enzyme is shown as lowering the hump and thus increasing the traffic over it. Make a highway out of this, with automobiles passing in either direction, and it is an interesting metaphor. However, it does not show how a catalyst can lower the hump. This can be done quite dramatically by means of the "shoelace" metaphor.

Imagine a man who is standing in a muddy field of indefi-nite extent who finds he must tie his shoelace. Now there is no danger of his flopping down into the mud (some-thing he does not want to do) as he stands there with his shoelace untied. Once his shoelace is tied, he is again in no danger. Both are stable positions.

During the process of tying his shoelace, however, he must either squat, bend, or raise his foot while remaining standing. In each of these alternatives, he significantly increases the risk that he will get part of himself muddy or that he may lose his footing altogether. He must therefore work very slowly and carefully through the unstable intermediate position.

If we imagine a whole series of men, all of whom must tie their shoelaces under similar conditions, one after the other, no one man starting till the one before has finished, the whole process will take a long time for completion, just because of the slowness of that unstable intermediate step.

Now supply a firmly placed chair and allow a man to sit in it. Once he is sitting, a foot can be raised without loss of stability. The shoelace can be tied without danger and the man can rise. The chair is not only a catalyst (again serving its purpose by offering a suitable surface), but it is one that specifically serves to stabilize the unstable intermediate position. It lowers the "energy hump" in a manner that can be clearly visualized.

Now a large number of individuals can tie their shoelaces, one after the other, much more quickly if the chair is used by each than otherwise. By stabilizing the unstable intermediate position, the chair-catalyst hastens the shoelace-tying reaction.

When a student is first introduced to enzymes, he meets those catalysts which are at once the most intimately involved with life and the most "mysterious." Yet, although protein in nature, enzymes share all the fundamental properties of catalysts generally. The brick-and-inclined-plane metaphor, the writing-board metaphor, and the shoelace metaphor all apply to enzymes as directly as to manganese dioxide.

But enzymes introduce additional refinements also. One way in which protein catalysts (enzymes) differ from the mineral catalysts is that the former are vastly more specific. It is not unusual to have an enzyme capable of catalyzing but one reaction out of the uncounted numbers possible. Yet this need not be accepted as a sample of the sweet mystery of life. Even a very superficial knowledge of protein structure would show that it is possible to build very complex surfaces out of protein molecules by varying the nature and arrangement of the amino-acid components. The value of a highly specialized surface over a generalized surface can be demonstrated by an extension of the shoelace metaphor.

A chair is a chair, but there are chairs and chairs. An ordinary kitchen chair is quite adequate as a catalyst with which to accelerate the typing of shoelaces. But now imagine a specially designed chair with back, arm rests, and foot rests that are motorized and capable of automated motion. As you sit down, your weight upon the seat closes a contact and up flies one of the foot rests, lifting your foot to just the right height. Simultaneously, the back moves forward, tilting you appropriately, while the arm rests move inward, bending your arms at the elbow and gently forcing your hands together. In a fraction of a second, and without effort on your own part, you have assumed the shoelace-tying position and a padded lever gently ejects you from the seat. It is now ready for another individual.

Obviously such a specially designed chair would hasten the shoelace-tying reaction to a greater extent than the generally designed kitchen chair could. It would also further stabilize that unstable intermediate position. Furthermore, such a specially designed chair by the very virtue of its specialization becomes less useful for other purposes. Unthinkingly, a young man might attempt to use it in order that he might hold his best girl on his lap. The motion of the various portions of the chair may surprise him. Yet even though he might find those motions endurable under the circumstances and even pleasant, he would almost certainly be disconcerted by the final ejection, as would the young lady also in question.

And if you only intended to use the chair to read a newspaper, you would abandon it in disgust even before you were ejected. In either case, you would seek a generally designed chair on the next occasion, or a chair specially designed for girl-holding or paper-reading as the case might be.

In short, the specially designed chair (enzyme) is at once a more efficient and more specific catalyst than the generally designed one (ordinary mineral), each characteristic almost necessarily implying the other.

Nor need we devise imaginary chairs to make the point.

One can as well, if not as ingeniously, use the notion of a barber's chair, a dentist's chair, a soda-fountain stool, or an electric chair and compare each with a kitchen chair to point out how a specialized surface at once increases efficiency and specificity.

The notion of specificity enters into the idea of competitive inhibition (see Chapter 2) as well. An enzyme may specifically catalyze the breakdown of substance A, let us say. It will not catalyze the breakdown of different substance B, nor yet of similar (but not identical) substance A', yet the presence of A' will interfere with the normal functioning of the enzyme with respect to A, while the presence of B will not.

Here we can use that most familiar of all enzyme metaphors, the "lock-and-key" metaphor. An enzyme working on a specific substance A may be compared with a lock to which A is the key. Substance B, which is nothing like A, is a key with its shaft completely different in grooving from that of A. It cannot even be inserted into the lock. As far as the lock is concerned, the presence of B has no meaning.

But now you have a substance A' which is similar to A. It represents a key with a shaft similar to that of A. Therefore A' can be inserted into the lock. However, the notches of A' are not similar to those of A. Therefore A' will not turn the lock. But it is *occupying* the lock. While it is there, though it will not turn, neither will it allow A to enter. The lock is temporarily useless or, if you prefer, the enzyme is inhibited.

The student will not only meet enzymes, he will meet groups of enzymes. A day will come when he will find that compounds within the body yield energy by having their hydrogen atoms transferred two at a time from compound to compound, till, at the end, they are attached to oxygen to form water. Most of the energy released in the process is stored in the form of compounds called "high-energy phosphate esters," about three of these being formed for each pair of hydrogen atoms transferred.

The hydrogen transfer from position to position is rather like a bucket brigade, with each step catalyzed by a separate enzyme.

Why the series of steps and the series of enzymes? Would

it not be better and simpler to combine the hydrogen atoms directly with molecular oxygen in a single step and use but a single enzyme to catalyze the reaction. As usual, we can find a metaphoric answer—the "staircase" metaphor.

Suppose it were necessary for a man to move from the fifth story to the ground floor and store the gravitational potential thus given off by using the energy of his downward motion to wind up three clocks. He could do this by pulling chains as he passed each clock and thus raising their weights by the pull of his own weight as he moved downward.

If he goes from fifth floor to ground level by means of five flights of stairs (a multi-enzyme system) he can, in the process, move relatively slowly, seize the clock chains surely, and pull them smoothly without breaking stride.

The man might also go from fifth floor to sidewalk level by jumping over the banister and down the stair well (the single-enzyme method). He would get to the sidewalk more simply and more quickly and lose gravitational potential as surely as by a stately progress down the stairs. However, he will find it difficult to snatch at the clock chains as he passes. He will release energy, but will not store any.

Again, the stair method of going from fifth floor to ground floor is reversible. One can move back up those same flights of stairs from ground level to fifth floor without a prohibitive expenditure of energy. However, having jumped down in a single bound, one cannot, alas (even supposing one to be in a position to try) bound back to the fifth floor in a single leap.

Similarly, a multi-enzyme reaction, in which each component step involves a relatively small energy change, allows a more efficient energy storage and is, at the same time, more easily reversible and hence more efficiently controlled by the body. The large energy change of the one-step method (although apparently this is the simpler alternative) makes it difficult to store energy efficiently and still more difficult to reverse matters at need.

These metaphors are not intended to be exhaustive, or even to be samples of the best possible; they are merely those which appeal to my own imagination. It is not the individual metaphors I value, but the principle of metaphor. Metaphor is itself a catalyst. By its mere presence and without actually increasing the scientific content of a course, it hastens the process of learning and is not used up thereby.

Chapter 8   A Pinch of Life

From the previous chapters in this book, one would be justified in deciding that life was a most subtle and complex phenomenon, the understanding of which taxes human ingenuity to the limit—and possibly beyond. Yet of what are living organisms composed that make possible this marvelous phenomenon?

If the human body, for instance, were broken up into separate atoms and the different kinds of atoms carefully segregated, two things would be obvious: (1) almost all the atoms would fall into a mere half-dozen varieties; (2) they'd be very common varieties.

In the first place, the body is mostly water, and each molecule of water is composed of two hydrogen atoms and an oxygen atom. Both hydrogen and oxygen atoms are found in most of the other molecules in the body as well. Aside from water, the body is made up mostly of organic (that is, carbon-containing) compounds. The most important organic compounds are the proteins, which contain nitrogen atoms, along with hydrogen, oxygen, and carbon.

The chief inorganic, or mineral, components of the body are the bones. The most common atoms in these, aside from those mentioned already, are calcium and phosphorus.

If we were now to count the various atoms, it would turn out that of every ten thousand atoms in the body there are:

> 6,300 hydrogen atoms
> 2,550 oxygen atoms
> 940 carbon atoms
> 140 nitrogen atoms
> 30 calcium atoms
> 21 phosphorus atoms
> 19 atoms of other varieties.

This appeared in *Science World*, March 5, 1957.

This is a most unglamorous list. Oxygen is the most common type of atom on earth. Carbon, calcium, and phosphorus are also among the dozen most common elements, in the earth's crust, at least. Most of the atoms that make up the oceans are hydrogen, and most of the atoms making up the atmosphere are nitrogen.

But let's put the top half-dozen varieties of atoms to one side. They are the staples out of which life is made. What about those 19 out of every ten thousand atoms that belong to other varieties? Why are they needed? If we've made 9,981/10,000 of the body with six elements, can't we let the remaining 19/10,000 go?

Apparently not. Nature is like a good cook, who knows that although a cake is made mostly of flour, milk, and eggs, it needs a pinch of this and that.

Let's see, then, what the other varieties of atoms are. Instead of counting the atoms in every ten thousand, let's count them in every million. If we do, we find that of every million atoms in the body, there are:

998,100 atoms of the types I've mentioned so far
    570 potassium atoms
    490 sulfur atoms
    410 sodium atoms
    260 chlorine atoms
    130 magnesium atoms
     38 iron atoms
      2 atoms of all other varieties.

There we have another half-dozen elements present in medium-sized pinches. Each is a common element that we couldn't do without.

Sulfur atoms occur as essential parts of almost every protein in the body, so we can't do without them.

Sodium, potassium, and chlorine are present as electrically charged atoms ("ions") dissolved in the body fluid. Sodium ion and potassium ion both carry a positive electric charge. Sodium ion is mostly found in the fluid outside the cells and potassium ion in the fluid inside the cells. Chlorine atoms carry a negative electric charge and are, in that form, called "chloride ions" (with a *d*). These chloride ions are found both within and without the cell, balancing some of the positive charge of both the sodium ion and the potassium ion.

These positive ions are responsible, among other things, for the electrical phenomena of the body. Shifts in the distribution of sodium ions and potassium ions inside and outside the nerve cells are responsible for the tiny electric currents that accompany nerve impulses. Without them, no nerve impulses and without nerve impulses, no life.

About half the magnesium in the body is in the bones. The rest occurs as positively charged ions in the body fluids. Magnesium is involved in the energy reactions of the body. Little packets of chemical energy are shifted from compound to compound, usually by means of the action of a substance known as adenosine triphosphate (ATP). Every reaction involving ATP requires the presence of a magnesium ion, which is thus necessary in energy-handling and is therefore essential to life.

The hemoglobin molecules in the blood contain four iron atoms each. Hemoglobin picks up oxygen molecules in the lungs and carries them to all body cells. It is the iron atoms in the molecule that do the actual carrying, so we can't do without iron.

If you consider hemoglobin and ATP, you will see why the body needs only a few atoms of certain elements. Each hemoglobin molecule carries four oxygen molecules from lungs to cell, then goes back for a new supply. In the same way, each molecule of ATP shifts one energy packet, then is reformed so it can go back for another.

Imagine bricklayers building a house. You don't need one bricklayer for each brick. One man can lay a million bricks if he works long enough. Though you need many bricks, you need only a few bricklayers.

In the same way, we need a great deal of oxygen, but only a small quantity of iron; or a great deal of energy, but only a small quantity of magnesium—just enough to help the hemoglobin and ATP, respectively.

Of course, we don't always know why one particular element, and no other, is needed for a certain job. Why magnesium ion, for instance? Why couldn't calcium ion (which is quite similar chemically) do the job with ATP? A good question, but so far no good answer.

The bricklayer reasoning applies to other essential elements needed in even smaller pinches than iron. These are the "trace elements."

If we count the atoms, not in every ten thousand or even million, but in every billion, we find that of every billion atoms in the body, there are:

999,998,000	atoms of the types mentioned so far
1,500	zinc atoms
170	manganese atoms
170	copper atoms
125	fluorine atoms
20	iodine atoms
10	molybdenum atoms
5	cobalt atoms.

Of these, fluorine is found almost entirely in the teeth, and is not really necessary to life, but only to healthy teeth. The other trace elements are essential to life.

Iodine atoms form part of the hormone molecules manufactured by the thyroid gland. The thyroid hormones control the rate at which the body produces and uses energy. It takes just a tiny bit of the hormone to do the job, just as it takes a tiny thermostat to control a huge furnace. The hormone won't do the job without iodine, so that element is essential to our body.

Of all the essential elements, iodine is the rarest in nature. Despite the small amount we need, it is sometimes present in insufficient quantities in the soil of many regions and therefore in the plant food grown in that soil and in the animal life that feeds on those plants. It is therefore necessary sometimes to add a pinch of it to a city's reservoir, or to make use of iodized salt (table salt to which traces of iodine-containing substances have been deliberately added).

Manganese, copper, zinc, molybdenum, and cobalt are each associated with some of the enzymes that are needed by the body to catalyze certain essential reactions (see previous chapter). It is because without them the enzymes won't work that they are essential.

You may wonder how the body can get any use out of an element like cobalt, when there are only five cobalt atoms out of every billion in the body.

But how little *is* five out of a billion? It is estimated that the human body contains some fifty trillion cells, but an atom is so much smaller than a cell that each cell, microscopic though it is, has room in it for at least a hundred trillion atoms.

If five out of every billion of these atoms are cobalt, then each cell can have an average of 500,000 cobalt atoms. Which shows that even the smallest pinch is not so small after all.

And now that we have the recipe of living tissue, what are the chances that we can take quantities of these atoms, put them together in the right proportion, and—

But that's the subject of the next chapter.

Chapter 9   Constructing a Man

In September of 1965, the chemists at the 150th National Meeting of the American Chemical Society were exhorted by their president, Dr. Charles C. Price, as follows:

"I would like to suggest a timely question of great public importance to which the scientific community and the Government should now be giving serious consideration: the setting of the synthesis of life as a national goal. . . .

"It seems to me we may be no further today from at least partial syntheses of living systems than we were in the 1920s from the release of nuclear energy—or in the 1940s from a man in space."

Imagine that! The synthesis of life! It is a theme as old as civilization.

In ancient times there were the girls of gold who (according to Homer) helped Hephaestos, Greek god of the forge, form the armor of Achilles. In medieval times, there was the tale of the golem, an automatonlike creature made of clay

A version of this article appeared under the title "Conceived in the Love Bed of Science" in *True*, February 1966. Copyright © 1965 by Fawcett Publications, Inc.

into which life was infused by Rabbi Löw of Prague through the use of the ineffable name of God. And in modern times there is the well-known story of Pinocchio, the wooden marionette who came to life.

Will the age-old dream ever become reality, or will it be nothing more than science fiction forever?

That same question was asked in 1960 at a gathering of scientists interested in the problem. Scientists are cautious individuals and some placed the synthesis of life thousands of years in the future; others, more daring, placed it centuries in the future; some wild optimists only decades.

But when Hermann J. Muller, the Nobel-prize-winning geneticist, was asked the question, he answered firmly, "Five years ago!"

Surely it sounds ridiculous to say that life was synthesized in 1955. What can Muller have meant?

Well, if Muller's words sound like a paradox, that paradox rests in the definition of life, and in how simple an object one is willing to call a living system.

The nonscientist, when he thinks of "life," tends to think of complicated systems indeed. To begin with, he is likely to think of man himself. If he visualizes the formation of synthetic life, he is apt to conjure up dim memories of Frankenstein. He may picture a cleverly fashioned artificial body of a man (or woman) lying upon an operating slab while the scientist pours "life" into it by way of some exotic radiation or some rare chemical.

Yet that, as sure as anything can be, is *not* how life will *ever* be created.

Why mold a human being, complete with flesh and bones, muscles and brain, glands and blood vessels? Nature doesn't —not all at once. No one begins life as a complete adult. All living organisms of any complexity at all, including human beings, are self-building devices that begin quite simply (at least, in comparison to the final product).

Living organisms are composed of cells, tiny (usually microscopically tiny) blobs of life. The human body is composed of some 50 trillion cells, but some very simple forms of life, such as the amoeba, are composed of single cells.

And even those organisms built up of many trillions of cells begin life as a single cell, the fertilized ovum. A man or a woman is formed, in actual fact, from a little blob of living jelly, a blob just barely big enough to see with the naked eye in a strong light. From that fertilized ovum, properly nourished in the female uterus by the mother's placenta, a baby containing about two trillion cells is formed in a period of nine months.

To create a man, then, it would be sufficient to create a fertilized ovum. Synthesizing the ovum is hard enough, goodness knows, but not nearly as hard as synthesizing a full-grown, perfectly formed man. Once the ovum is formed, it can carry on from there. To be sure, it would have to be nourished adequately thereafter, but we are approaching the ability to do that.

Biologists can keep isolated organs, and even scraps of tissue, alive for considerable lengths of time. Before World War II, the well-known surgeon, Alexis Carrel, managed to keep a scrap of embryonic chicken heart alive and growing (it had to be periodically trimmed) for over 32 years. This was quite a feat, for special precautions had to be taken to keep the tissue from becoming infected by bacteria. Nowadays, with the development of antibiotics, infection is no longer a serious problem and tissues can be kept alive more easily.

As for the fertilized ovum, progress has been made there, too. It is quite within the realm of present technology to transfer a fertilized ovum into another body and let it develop there. This was done as long as 70 years ago with rabbits. It has been done with most common laboratory and farm animals, and, if the species are the same, a foreign womb often produces normal young. One prize ewe has given proxy birth to 11 lambs in one season instead of the usual one or two.

What stops this from being done with human beings is distaste rather than inability. In 1961, Dr. Danielle Petrucci of Bologna, Italy, claims to have taken an unfertilized female human ovum out of the ovary, had it fertilized in an artificial glass "womb" in which the embryo lived and grew for some time.

The suggestion is often made that the sperm cells of a

remarkable man might be frozen and kept alive in order that his genes could be passed on to many more offspring than he could produce in an ordinary lifetime. Similarly, if fertilized ova can be passed on to other women for "brooding" purposes, the genetic mother could produce 13 potential offspring a year. A robust, young "brooder" might actually produce healthier young than the supplier of the ovum, but would not herself contribute genetically to the young to whom, in this manner, she would give birth.

Do we need a human "brooder" at all, for that matter? Suppose the ova banks and sperm banks are used as sources for the necessary cells, which would be allowed to undergo fertilization and subsequent development within a synthetic womb. Such a womb would reproduce necessary environmental conditions such as temperature and pressure, perhaps even the sound and vibration of a simulated maternal heartbeat.

Till now, fertilized eggs have been developed outside the body only through the very early stages. Before actual organs begin to form, the process stops. If the equivalent of a placenta could be designed, then the way would be clear for the artificial development of a complete human being from an egg cell and a sperm cell. There is a name for such a process: "ectogenesis."

Ectogenetic development would clearly be of great scientific value for it would enable us to learn a great deal about the development of life through continuous observation.

A society in which ectogenesis becomes common might consider it desirable to develop embryos under ideal conditions, safe from the ills, shocks, malnutritions, and accidents to which a real mother might be subjected.

In an overpopulated world, it is becoming important to find techniques for maintaining the population at a controlled level. With ectogeneria, this needed goal of population control would be easy. (And their existence would not interfere with ordinary sexual activity which, after all, is not always intended for the production of children anyway.)

Furthermore, it would be possible to study the embryos closely and bring to fruition only those which tested out as free of serious physical or biochemical abnormalities, something that cannot be done while the baby is invisible and unreachable in a human womb.

The complete divorce of childbearing from the sex act would undoubtedly revolutionize the general attitude of mankind toward sex. By removing the false extremes of good and evil from its image, sex might at last become the natural function it really is, and a potent source of neuroses may be removed forever from the human race.

Naturally, there are anti-Utopian aspects of such a possible future. Who is to decide who shall qualify as potential parents? What shall be used as a basis for qualification? Actually, we don't know enough right now to be safe with an ectogenetic society. But we can hope that by the time our science and technology has advanced to the point where such a society is possible, we *will* know enough.

But an ectogenetic society does not fulfill the dream of created life. It is not enough to take life which already exists —in the form of a fertilized ovum—and chivy it on to fruition. In that way, we are only doing in glass what the human body is doing in flesh.

How about actually forming a cell to begin with out of nonliving materials? In that way we could form a completely new specimen of life, which would owe nothing at all to previous life.

Easily said—but not easily done. Even a single cell is a vastly complex system, far more complicated, despite its size, than the giant ocean liners and skyscrapers that man *can* construct.

We might turn to nature and ask how a cell is formed at her hands. The answer is simple. All cells that exist today have been formed from other cells. All your cells are formed from the original fertilized ovum which was your beginning. That was formed out of a paternal sperm cell and a maternal egg cell, and those were in turn formed out of other cells which can be traced back to the fertilized ova out of which your father and mother formed—and so on back and back and back, for all creatures now alive, for billions of years.

But far back at the very beginning, cells must have formed

84

from noncells, and how did *that* happen? We don't know. We can only make reasonable guesses.

It took a very daring mental leap for scientists to begin to suspect that the passage from noncells to cells, from non-life to life, might have taken place as a matter of blind, random, chemical processes. Our Western culture has been too imbued with that sacredness and uniqueness of life to make it a random product, too ingrained with the notion of the divine, purposeful creation of life and man, as described in the Bible. Even the rejection of the Bible's story of creation couldn't get rid of the haunting whisper within.

It is perhaps no accident that the spell was broken by a biochemist of the Soviet Union, which is officially atheist in its philosophy of life. This biochemist was A. I. Oparin, who began to write on the subject in 1924 and who felt that cells would arise through inevitable, and rather simple, natural phenomena.

He considered, for instance, the natural formation of droplets of one kind of liquid suspended in another, under conditions prevailing in the primordial ocean.

Going much farther in this direction now, over a generation later, is Sidney W. Fox of the Institute of Molecular Evolution at the University of Miami.

Professor Fox begins with a chemical system designed to represent the conditions as chemists believe them to have been on the primordial earth of several billions of years ago and subjects the system to heat—of which there was always plenty, thanks to the sun.

Starting with simple compounds of the type that would have been common eons ago, he finds that heat alone will suffice to form amino acids, and then force these together in long chains to produce proteinlike compounds which he calls "proteinoids."

This worked best at temperatures above the boiling point of water and some biologists doubted that such a process could take place on the primordial earth without the proteinoids breaking down as fast as they are built up. Fox, however, draws a picture of proteinoids forming on hot volcanic ash and then being dissolved and washed away by hot rain before much of it has had a chance to break down.

Fox found that when his proteinoids were dissolved in hot water and the solution was then allowed to cool, the large proteinlike molecules tended to collect and clump out in the forms of little globes he called "microspheres."

These microspheres resemble very simple cells in some ways. They are like small bacteria in size and shape. They are surrounded by a sort of membrane as cells are. They can be made to swell and shrink by appropriate changes in the surrounding fluid, as cells can. They can produce buds, which seem to grow larger sometimes and break off. They can divide in two, or cling together in chains. The material within the microspheres even display some of the properties reminiscent of the workhorses of living tissue—the enzymes.

The microspheres cannot be considered alive by any ordinary standard, but can one really speak of life and nonlife as being separated by a sharp boundary? Many biologists think not. Life and nonlife are separated, rather, by a broad zone within which objects may be regarded as progressively more alive and less nonalive. If so, the microspheres, while a long way from the completely alive side of the boundary zone, are at least a small way past the nonalive side.

It may be that Fox, and others, may push the microspheres farther and gradually approach and pass the boundary of undoubted life. And maybe not. It is hard to tell.

Perhaps it is a mistake to try to jump from nothing to the cell. It might well be that the cell is not a suitable object as the immediate goal of the life-synthesizers. It is very likely that it was not the first product of the natural evolution of life. The cell, as we know it today, may not be an example of primitive life at all, but is rather the end product of a long period of evolution. For uncounted millions of years before the first cell arose, there must have been simpler structures in existence. Once cells were formed, however, their superior efficiency drove these simpler "pre-cells" out of existence and leaves us today with a world of life in which cells *seem* the simple beginning only because they have killed off the competition.

But the "pre-cells" have not gone without leaving any trace at all.

Within every cell are smaller bodies. There is the cell

nucleus, for instance, which contains chromosomes that control the machinery of inheritance. Outside the nucleus are mitochondria which contain the energy-handling apparatus. In plant cells are chloroplasts which are living versions of the solar battery, equipped to convert the energy of sunlight into the chemical energy of stored food.

All these "organelles" may represent the remains of primitive "pre-cells." Such pre-cells may finally have come to exist in co-operation, forming complex structures much more efficient than themselves taken singly. These pre-cell co-operatives (what we now call cells) then took over the world.

Of these organelles, the most fundamental seem to be the chromosomes. Every species has a certain characteristic number of these present in each cell. Each human cell has 46—resembling blunt, thick, intertwined spaghetti strands at certain stages of the growth of the cell.

Each time a cell divides into two cells, each chromosome undergoes changes that produce two chromosomes that are each replicas of the original. The process is called "replication." If we trace back the 46 chromosomes in each of the 50 trillion cells in the human adult, we find them originating from the 46 chromosomes of the original fertilized ovum. The chromosomes of the fertilized ovum were obtained from the two parents, half from the father's sperm cell, half from the mother's egg cell. These can be traced back to the fertilized ova out of which the parents originated and so on.

It is the chromosomes that supervise the formation of enzymes within the cell. In every generation, chromosomes from two parents form a new combination; and besides, minor changes are always taking place in chromosomes as one passes from parents to children. As a result no two individuals (barring identical twins who arise from the same fertilized ovum) have precisely the same chromosomes and no two individuals form precisely the same enzymes.

It is the enzymes that supervise the chemical functioning of each cell and that thus lend each creature its life and individuality. We might therefore view the chromosomes as the real beginning of the cell, just as we view the cell (in the form of the fertilized ovum) as the real beginning of the complex adult.

That, perhaps, is the essential component yet missing from

Fox's microspheres. If we could synthesize chromosomes and place them within the microspheres, we might finally have indisputable life. Or, perhaps, if we form chromosomes, we can encourage them to form their own cells.

This may be so, for there is actual evidence (aside from mere reasoning) that the chromosomes are more fundamental than the cell. Cells do not exist without chromosomes; but chromosomes (after a fashion) exist without cells.

These objects, which resemble bare chromosomes, are what we call "viruses." These are much tinier than the cell and much simpler in structure. They are of the size of chromosomes and in chemical structure and function resemble chromosomes.

Viruslike objects may have existed billions of years ago, before the evolution of cells, may have been capable of independent reproduction. They may have had within themselves all the capacity for growth and multiplication, and may therefore have been somewhat more complex than modern viruses.

For the viruses that exist today have been spoiled by the very availability of cells. The modern virus is a complete parasite that has shed the equipment it needed for independent life and merely maintains itself, and no more, outside the cell. Once it gets a chance to enter a cell of the proper type, however, it can make use of the *cell's* chemical machinery for its own purposes; multiplying itself at the expense of the cell's own needs and sometimes killing its host in the process.

There was some doubt at first as to whether the virus ought to be considered as living, but most biologists have now decided in favor of characterizing the virus as alive. It is this, in part, which gives rise to the source of disagreement among scientists as to when life may be synthesized. If by life one means complex cells, then synthetic life may be a long way off. If, however, one considers a virus to be alive, then the goal is much closer than we might think.

Ordinarily, for instance, a virus only reproduces itself within cells, making use of the necessary enzymes, raw materials, and energy sources present in such abundance there. But suppose we take a small quantity of virus and supply it with the necessary work materials *outside* the cell.

In October 1965, Professor Sol Spiegelman of the University of Illinois reported on his work in this direction. He had managed to produce virus in the test tube. In a sense this represents a synthesis of the very simplest form of life, but in a truer sense it is not a complete synthesis. A bit of virus had to be used as a starter, so that the process rather resembles that of growing a chicken (or a human being) from an egg. What we would like to see is *completely* synthetic life; life formed out of a system containing no life whatever to begin with.

To visualize the possibilities there, let's look more closely at the chemical structure of a chromosome or virus.

The interior of a chromosome, or of a virus, is made up of a long coiled chain of atoms forming a molecule of nucleic acid. The particular variety of nucleic acid in chromosomes, and in the more complex viruses, is "deoxyribonucleic acid" usually abbreviated DNA. Surrounding the DNA is a coating of protein.

The molecules of both DNA and protein are exceedingly complex and have within them an extraordinary capacity for variability (see Chapter 2). Biochemists had been aware of the versatility of proteins for a century and more, whereas nucleic acids were relative late-comers on the biological consciousness. Furthermore, proteins are built up of some twenty different types of units, whereas nucleic acids are built up of only four. It was taken for granted, therefore, up to the mid-1940s, that it was the proteins, not the DNA, that were the key chemical in the chromosome or virus. Beginning in 1944, evidence began to pile up in astonishing fashion in favor of DNA.

As an example of a dramatic experiment, there is the one conducted in 1955 by Heinz Fraenkel-Conrat, a research biochemist at the University of California, at Berkeley. Fraenkel-Conrat managed to separate the protein coating and the nucleic acid core of a virus. Separated in that fashion, neither the coating alone nor the core alone could infect cells. The virus seemed dead. He then mixed the coats and cores and some of these came together again to form complete viruses capable of infecting cells.

For a time it seemed that a living organism had been

killed, then restored to life. Even though the organism concerned was at the simplest possible level of life, this feat made headlines.

However, it turned out that life had neither been killed nor restored. The nucleic acid core had life in itself. Every once in a long while it managed to invade a cell without the presence of its protein coating. The protein helps the nucleic acid get into cells (as a car helps a man get from New York to Chicago) but the nucleic acid can manage it—with difficulty—alone, just as a man can walk from New York to Chicago if he absolutely has to.

And it was shown that when an intact virus invades a cell, it is only the nucleic acid core that does so. The protein coating, having fulfilled its task of facilitating the entry, remains outside. Within the cell, the nucleic acid core not only replicates itself but also supervises the formation of a protein coating (a protein not quite like any of the proteins the cell would form of its own accord).

Scientists had begun to concentrate on the nucleic acid molecule after 1944, and particularly on DNA, its most important variety. A New Zealand-born physicist, Maurice H. F. Wilkins, who had been one of the British scientists working on the atomic bomb during World War II, studied DNA by bouncing X-rays off their molecules. The photographs he produced were studied by a British colleague, biochemist Francis H. C. Crick, and his American co-worker, Dr. James D. Watson (who, in his youth, had been one of radio's Quiz Kids). In 1953 they worked out the structure of DNA, showed it to be a double string of four different, but closely related, units called "nucleotides."

There are uncounted numbers of possible patterns to the DNA molecule, according to the order in which the different units are distributed. Watson and Crick showed how the molecule could behave so as to form new molecules with exactly the same pattern.

Other biochemists painstakingly worked out the manner in which the DNA pattern was transferred to the analogous pattern of a protein, so that specific portions of the DNA molecule produced specific enzymes and thus controlled the chemistry of the cell. The transfer of "instructions" from the nucleic acid pattern to the enzyme pattern is called "the genetic code."

Apparently, then, the basic chemical reaction of life is the ability of the DNA molecule to replicate itself. That is the whole of the law; all else is commentary. Therefore, if we are capable of forming a DNA molecule from simple, nonliving substances, we have synthesized the very beginning of life. There would be an unfathomed distance between this and the synthesis of a man, perhaps, but it would be a true beginning. We would have crossed the threshold between nonlife and life.

How did nature itself pass this threshold? The threshold must have been passed billions of years ago, when there were no enzymes to do the work, and no nucleic acids to serve as blueprints.

It would appear that on the primordial, lifeless earth, only certain simple molecules could have been present in any quantity in the ocean, where life is usually thought of as having originated, and in the atmosphere. The nature of these molecules can be deduced from the overall composition of the early earth (based upon the known composition of the sun and of the universe generally) and upon the known laws of chemical combination.

Suppose we begin with such molecules—water, ammonia, methane, hydrogen cyanide, and so on—and add to them energy in the form of ultraviolet light, radioactivity, electron streams or lightning (all of which would have been available on the primordial earth). What would happen?

Charles Darwin, the founder of the theory of evolution by natural selection, had considered this question a hundred years ago and wondered if the chemicals of living tissues might not be built up out of such a system; if there might not have been "chemical evolution" as well as the evolution of species.

The first to try to investigate the matter experimentally was Melvin Calvin at the University of California. In 1951, he began to note the effect of energetic radiation in building up complex compounds out of simple ones.

In 1952, Stanley L. Miller at the University of Chicago went further. He placed simple chemicals of the type present on the primordial earth in a container absolutely free of living matter and subjected them to the action of an electric discharge for a week. When he was done, he detected the presence of a number of more complex substances than he had started with—including four different amino acids, each one a variety that was present among the units of naturally occurring proteins.

Since then, a number of other chemists, such as Philip H. Abelson at the Carnegie Institution and Joan Oro at the University of Houston, have experimented in similar fashion. Under the impact of various forms of energy, complex compounds were formed out of simple starting materials. Then, using those complex compounds as new starting material, still more complicated compounds were built up. All the compounds that appeared were similar to the key components of living tissue. The natural route followed by this blind, random buildup seemed always to point directly toward life.

In particular, a Ceylonese-American biochemist, Cyril Ponnamperuma, in work at NASA's Ames Research Center, demonstrated the production of portions of nucleotide molecules, the building blocks of nucleic acids. The complete nucleotide contains atoms of phosphorus. Therefore, simple phosphorus-containing substances were added to the mixture being worked with. Along with Carl Sagan and Ruth Mariner, Ponnamperuma engaged in a course of experimentation that ended up with the production of a complete nucleotide molecule. By 1963, the nucleotides had been formed in the particular high-energy form which could be used to produce the nucleic acids themselves.

Indeed, in September of 1965, Ponnamperuma announced that he had progressed another step. He had succeeded in forcing two nucleotides to join into a "dinucleotide," one that contained the same kind of linkage that joins nucleotides in natural nucleic acids.

Clearly, then, scientists have a smooth chain of synthesis stretching from the simple compounds that existed on earth when our planet first took its present shape, clear up to molecules that point directly at nucleic acids. There are no gaps in the chain.

One gets the picture of inevitable changes up through

the molecular level. Start with a planet like the earth, with a complement of simple compounds bound to exist upon it, add the energy of a nearby sun, and you are bound to end with nucleic acids. You can't avoid it, and all that scientists need do is guide the process and hurry it up.

The synthesis of nucleotides by convenient chemical methods (not necessarily like the random processes that take place in the system worked with by Ponnamperuma) are now old stuff. The Scottish chemist, Alexander R. Todd (now Baron Todd of Trumpington), had synthesized the various nucleotides in the 1940s.

But what about the passage from nucleotides to nucleic acids themselves?

In 1955, the Spanish-American biochemist, Severo Ochoa, at New York University, began with a solution of nucleotides in high-energy form and with appropriate enzymes and formed molecules closely resembling natural nucleic acids— even though there hadn't been a single nucleic acid molecule present in the mixture to serve as a model.

It is this synthesis of nucleic acid from simple molecules that Muller must have referred to in 1960, when he said that life had been synthesized five years before.

To be sure, nucleic acid molecules that are synthesized without a blueprint are put together in random fashion and tend to be simpler than the natural molecule. Such synthetic nucleic acids don't fit the workings of any particular cell and cannot enter cells and multiply there. They may possess the potentiality of life but they can't be made to demonstrate that potentiality in action.

The biologist is now at the stage where he can:

1) Form nucleic acid molecules modeled on some natural molecule present in the system. Such molecules may be considered as alive but are not formed out of completely non-living starting materials.

2) Form nucleic acid molecules out of completely non-living starting materials. Such molecules cannot so far be made to demonstrate the phenomena associated with life.

To form an indubitably living nucleic acid molecule out of completely nonliving starting materials is still beyond the powers of science—but surely not for long, and that is what

Price meant in the quotation with which I started this chapter.

Let us look forward to the possible consequences that will follow when mankind is able to form synthetic nucleic acids, synthetic viruses, synthetic chromosomes, synthetic life.

Are there immediate dangers? Suppose scientists manufacture a new virus that *can* invade a cell; a new virus against which man, perhaps, has never developed any defenses. Might a new, unimaginably deadly plague spring out of the test tube to wipe out humanity and, pehaps, all cellular life?

The chances of this, actually, are small indeed. The invasion and exploitation of a cell by a virus is an extraordinarily complex phenomenon. That it works at all is the result of billions of years of slow evolution, and a virus is usually adapted to parasitize only certain cells of certain species.

To suppose that a virus will be formed that, just by accident, will happen to fit all the idiosyncrasies of some types of human cell, and possess the capacity of destroying them is to ask entirely too much of chance. It may not be mathematically impossible, but it is wildly improbable.

Let us turn, then, to more constructive and optimistic possibilities.

The day is dawning now, perhaps, when we may be able to duplicate an early triumph of mankind at a far more subtle and sophisticated level.

Once, in dim prehistoric times, man was a food gatherer. He ate wild animals that he could kill or fruits and berries that he could pick. If he was unlucky at the chase or at berry gathering, he went hungry.

Then came the time when mankind learned to tame animals, feed them, and watch over them, making use of their milk, wool, and labor, and slaughtering them for food whenever necessary. He also learned to sow plants and harvest them. From a food gatherer, he became a herdsman and farmer, and much more food became available. Mankind had its first population explosion as a result of these discoveries about 10,000 years ago.

As far as the substances of the cell are concerned, we

are still in the food gathering stage. Take insulin, for instance. Insulin is a protein produced by certain cells in a gland called the pancreas. It is not an enzyme but a hormone that is necessary to the proper function of the body. In its absence (or short supply), diabetes results (see Chapter 3).

A man with diabetes can live a normal life if he receives regular injections of insulin. Such insulin is obtained from the pancreases of slaughtered cattle and swine. We "gather" the insulin from the pancreases we happen to come across—exactly one per slaughtered animal. This means the supply is limited.

The supply happens to be enough, but why gather insulin if the possibility comes that we may obtain it from "herds" of molecules? Suppose we don't filch the insulin from the pancreas cell but filch, instead, the nucleic acid molecule that brings about the formation of insulin. If we "herd" the nucleic acid, keeping it supplied with the raw materials it needs, it can form insulin in indefinite quantities, as a cow produces milk. We will then have our own supply of insulin and won't have to depend on the number of animals we happen to slaughter. Furthermore, we can make the nucleic acid form replicas of itself, perhaps, and never have to go back to the animal even for that.

Can we see a future in which factories are built where the working machinery consists of submicroscopic nucleic acids? Might not mankind gather a repertoire of hundreds or even thousands of complex enzymes and other proteins? Some of the enzymes could be used to bring about chemical reactions more conveniently than any methods now used. Others might be used in medicines or in helping to construct life.

It is even possible that some of the material formed might serve as food. The manufactured protein could be used to fortify natural foods in the undernourished parts of the globe. It would be expensive, particularly at first, but it would consist of pure digestible substance, free of bone, gristle, and fat, and of particularly high nutritive value.

The average man on earth might be expected to resist the introduction of such "unnatural" items into the diet, but how about colonies on the moon or on Mars? In the absence of cattle and apple trees on those worlds, and considering the expense of carting food across space, it might be just the thing to bring hard-working nucleic acids along. The raw materials for the molecules could be built up very largely out of minerals found on the spot. (Limestone and hydrated silicates would be very helpful.)

In fact, it may not be until nucleic acid molecules are properly harnessed that the colonization of the solar system will become a fully practical venture.

Nor need mankind confine itself to following the feats of the cell with complete slavishness. Nucleic acids do not, after all, always produce *exact* replicas of themselves. Sometimes, small errors are introduced into the replica. This is not an entirely bad thing, for occasionally the errors result in a new kind of nucleic acid that is useful to the cell in which it occurs. It is these random changes in nucleic acids that have resulted in the process of evolution and in the long, two-billion-year-or-more development that has made a man out of an amoeba.

Men can encourage the appearance of such changes in nucleic acids during replication. By treating nucleic acids with heat, radiation, or certain chemicals, the number of errors is made to increase. The new nucleic acids form protein molecules (many of which are enzymes) that are also in error, that have patterns somewhat different from the original pattern. Most of these new proteins may well be useless. A few, however, might have new and important properties not met within nature.

(Chemists have gone through this process before. A hundred years ago, they learned to put together chemicals that are not found in nature. In so doing, they discovered new dyes, new medicines, and even new giant molecules, such as those in synthetic fibers and plastics. In many cases, the new substances were actually improvements on nature in certain ways.)

Might we not, then, form new nucleic acids which will form new proteins that will turn out to be improvements on nature in one way or another? In addition to "herding" our nucleic acids, we will "breed" new varieties, just as we now breed new varieties of cattle and wheat.

And can the new nucleic acid technology ever be applied to human beings directly? Let's speculate further.

Each chromosome is made up of hundreds, if not thousands, of nucleic acid units, each capable of bringing about the formation of particular proteins. The oldest name for these units is "genes." Each human being has his own set of genes, and every one of us, probably, includes in his own cells certain defective genes, incapable of forming certain enzymes in appropriate form.

Often this lack is not serious; sometimes it is. Scientists are learning to identify the genes by various techniques. In 1962, Robert S. Edgar, of the California Institute of Technology, identified about half the genes present in a particular virus, working out the nature of the enzyme each produced.

Eventually, given a set of chromosomes in a cell, techniques may be evolved that will determine the nature of each gene present. All cells in a given individual have the same set of genes, so that such a "gene analysis" can be made from the white cells in a drop of blood and the whole process will cost only a pinprick.

Perhaps the time will come when each individual will undergo such an analysis at birth. And once the set of genes is analyzed and identified, is there anything that can be done about it? Perhaps. From this chart of his defective genes the future state of his health may be predicted, preventive measures may be taken; his career may be planned with his physical potentialities in mind. The gene analysis card may become an essential part of a man—to be on his person at all times and on file at some central bureau as well.

Even though each cell in an individual has the same set of genes, the genes don't express themselves the same way everywhere. Cells specialize; some become nerve cells, some muscle cells, some skin cells, some liver cells, and so on. Each cell has its own set of enzymes, which means that in each kind of cell some genes are prevented from working, while others are encouraged to work at double time.

Exactly what it is that blocks some genes and encourages others, scientists do not yet know; but as of now, this is the most urgent problem facing biochemists, and they are working at it from several angles. Some are checking the proteins contained in the chromosomes; they may be the blocking agent. Others are studying the products of enzyme action; these very products, as they build up, may slow down the action of the enzymes that produce them. This "feed-back" may be involved in gene blocking. And, of course, still others are checking additional possibilities.

Suppose we learn enough to be able to unblock genes. In that case, we would have cells with all the capabilities of the original fertilized ovum. If the stump of an amputated arm or leg can be "despecialized" in this way, can it then be treated so as to cause it to grow back to a complete arm or leg? Can nerves be regenerated so that paralysis becomes a thing of the past; eyes rebuilt so that blindness becomes but a memory?

Or let us go back further and bring gene analysis to the original fertilized ovum. Suppose that a fertilized ovum is allowed to divide in two and that one of the new cells is then detached and removed. No damage is done, for the remaining cell can then proceed to divide again and again and produce a complete individual. (In fact, identical twins are born when the first pair of cells formed by the dividing fertilized ovum happens to separate, each cell going its own way.)

The cell removed can be used for gene analysis. It might then be possible to tell at the very beginning whether to allow the remaining cell to develop to babyhood or not.

Suppose, though, we find that a key gene in the fertilized ovum is defective but that otherwise the pattern is a very good one which will give rise to a superior human being. It would be a shame to lose that possibility for the sake of one gene. Could a healthy gene be substituted from some "gene bank"?

In 1964, Muriel Roger at the Rockefeller University reported having transferred an individual gene from one bacterial cell to another. The cell that received the gene could then produce a new enzyme it could not have produced earlier. So the idea of gene transferrals is by no means an inconceivable one.

Then, too, suppose a fertilized ovum has several defective genes, too many to salvage it for a complete individual. It might happen, however, that none of these defective genes would hamper the working of a heart, or a kidney. Could the ovum then have various genes blocked so that it will specialize at once and develop into only a heart or a kidney? Could we then have a supply of strong, young organs for use in transplantation?

This all sounds wild indeed, to be sure, but things are moving terribly quickly. Enormous, even undreamed-of, progress can be made in mere decades. Sixty years after the Wright Brothers' first fumbling plane flight, jet planes were circling the earth. Forty years after Robert H. Goddard had sent the first liquid-powered rocket 184 feet into the air, rockets were flying past Mars.

Who knows, then, at what stage of bio-engineering we will have come to rest by 2000 A.D.—a time many of us will live to see.

The capacity for bio-engineering is not something to look toward without a certain apprehension, of course. Will we know enough to play God with life and living things?

Perhaps not, but at least it won't be the first time man has taken the risk. He has been playing God ever since he began to apply his intelligence to the changing of environment. When man domesticated animals, invented agriculture, and built cities, he created "civilization." This altered his way of life profoundly and introduced problems that had not existed before. Yet on the whole, it represented an improvement, and we would not want to return to barbarism.

Again, when man built the steam engine, tamed the electric current, designed the internal combustion engine, and devised the nuclear bomb, he created a technology that wrenched his way of life from its moorings. Heaven knows, enormous problems were created, and yet again, few of us really want to return to a pre-industrial existence.

No doubt, an era of bio-engineering would introduce still another set of crucial changes and back-breaking problems, but judging from the past, man will, in the balance, manage. The benefits will outweigh the disasters.

Then, too, if man can really get started in the program of working out improvements for himself, it will be improved-man that will work on still further improvements.

Each accomplishment will make easier the next and, in the grip of this upward-tending spiral, mankind may achieve sanity at last, and finally emerge into the sun-drenched uplands of the human potential.

PART I
(continued)

CONCERNING THE MORE OR LESS KNOWN

B
NONLIFE

Chapter 10   The Flaming Element

From the moment of its discovery, the inflammable gas, hydrogen, has had a revolutionary effect upon mankind. It has broken down old theories and helped establish new ones. On two different occasions, it led men upward toward the stars. Now it points onward toward endless energy stores for man's future needs.

Its history began in flames, for in the 17th century, the early chemists produced a new "air" from iron and acid, an "air" which exploded when heated. They called it "inflammable air."

The English chemist, Henry Cavendish, who studied the new substance in 1766, found it produced something more remarkable than flame. When this gas burnt and combined with something in the air (oxygen, as it later turned out), drops of liquid were formed which proved to be water. Out of flame had come water.

The world of chemistry was astounded. For thousands of years it had been believed that water was an element; that it could not be formed from simpler materials. Now the combination of two gases produced water.

Inflammable air was given a new name, "hydrogen," meaning, in Greek, "water-producer." The formation of water from hydrogen was one of the clues that enabled the French scientist, Antoine-Laurent Lavoisier, to sweep away old theories and to establish modern chemistry.

But hydrogen was a wonder gas in more than one way. Not only did it flame and form water, but it was also incredibly light. A cubic foot of ordinary air weighs only 1¼ ounces. This is little enough, but a cubic foot of hydrogen weighs less than 1/10 of an ounce. In fact, hydrogen is the lightest substance known.

In 1783, the Montgolfier brothers in France had filled a silk bag with hot air and set it flying upward. The hot air was lighter than cold air and floated upward through the

A version of this article appeared in *Petroleum Today*, Winter 1961–62.

atmosphere as a chip of wood would move upward through water. When the hot air cooled, the silk bag (the first balloon) came down.

But why use hot air? The new gas, hydrogen, was much, much lighter than air even when it was cool. Its more powerful lifting force would carry a gondola aloft—and men inside the gondola.

All over Europe and America in the first years of the 19th century, hydrogen-filled balloons were drifting across the heavens. For some, it was merely a thrill, an exciting adventure. For scientists, it was a new way of studying the heights of the air—the first step toward the stars.

It could also mean commercial travel if only the balloons could be made independent of the wind. In 1900, the German inventor, Count von Zeppelin, built cigar-shaped balloons in aluminum frameworks and added a motor-driven propeller. The dirigible balloon (or "Zeppelin") was a ship of the air, borne aloft on the wings of hydrogen.

But hydrogen, for evil as well as for good, is a creature of flame. The gigantic bag of hydrogen was a container of explosive, an unmissable target for the enemy. And the enemy was, sometimes, nothing more than a spark of static electricity. In 1937, the hydrogen bag of the great dirigible, *Hindenburg*, burst into flame. In minutes, it was destroyed.

But the dirigible had had its day, by then, in any case. The future lay with heavier-than-air machines—smaller, swifter, and more capable of withstanding bad weather.

It looked as though hydrogen might be confied to earth-bound uses. It was used by chemists to reduce or "hydrogenate" organic materials in a thousand ways—turning inedible vegetable oils into useful solid shortenings, for instance. The flame of hydrogen was used by industry in the form of oxy-hydrogen torches which cut through steel as though it were cheese.

But beyond that?

Hydrogen was not defeated, however. If the dirigible went down in flames, the rocket went *up* in flames. And even as the last dirigible died, the day of the rocket was dawning.

Ordinary aircraft can maneuver only in air that contains a sufficiently concentrated supply of oxygen to burn the fuel in the engines. The air, furthermore, must be dense enough to support the weight of the machines.

A rocket, however, carries both fuel and oxygen. The two combine in white-hot fury, sending a blast of heated exhaust gases downward. Because part of the content of the rocket, in the form of these gases, is hurled downward, the rest of the rocket moves upward. (This is in response to the "law of action and reaction," or "the third law of motion," first expounded by the English scientist, Isaac Newton, in 1683.)

As the exhaust gases continue to stream downward, the rocket moves up faster and faster. Eventually, it will streak far above the atmosphere (which it does not need either for support or to keep its flame burning) and will penetrate outer space.

The height to which a rocket will rise depends, in part, upon the manner in which the exhaust gases are ejected. The more rapidly they jet downward (the more violent the "action"), the greater the velocity and altitude attained by the rocket (the more violent the "reaction"). Rocket scientists had to find the fuel that would produce the greatest upward reaction.

The earliest rockets, such as the toys used on the Fourth of July, and the scarcely-more-than-toys employed in 19th century warfare (the "rockets' red glare" of our National Anthem) used gunpowder. Gunpowder contains an oxygen-rich compound called "saltpeter." It also contains carbon and sulfur which, when heated, combine violently with the oxygen in saltpeter. Gunpowder is thus fuel and oxygen combined.

But gunpowder is not very powerful. In 1926, the American inventor, Robert H. Goddard, realized that much better could be done with liquids. On March 16 of that year, at his Aunt Effie's farm at Auburn, Massachusetts, he launched the world's first liquid-propellant rocket. His fuel, a mixture of gasoline and liquid oxygen, yielded about five times as much energy, pound for pound, as TNT. The great energy of this combination was soon sending rockets miles into the air at supersonic speeds.

Although an American fathered the modern rocket, it came of age under the Germans, who built the V-2 rockets during World War II. We brought some of these Germans to the United States in 1946 and got seriously to work. (Sadly, Goddard had died the year before.)

The gasoline-oxygen combination continued to be used, but it by no means represented an upper limit of energy potential. Of all chemical fuels, hydrogen (in combination with oxygen or fluorine) flamed most energetically. A hydrogen-powered rocket would rise much higher and lift a heavier load than would one powered by the same weight of gasoline or any other fuel.

Once again hydrogen seemed to be on the brink of an aerial career, but there was a catch. Hydrogen couldn't be used in its ordinary form. A pound of hydrogen takes up a hundred sixty cubic feet of space and the one thing a rocket lacks is roominess.

Hydrogen had to be obtained in compact form. It could be compressed under large pressures, but that was difficult —and dangerous. However, there is one way of compacting a gas without large pressure; cool it down into a liquid.

Nor was it rocketry alone that needed hydrogen compact and in quantity in those days after World War II. —A new bomb was being devised.

The ordinary uranium-fission bomb (the dreadful "A-bomb" that had ended Japanese resistance) was becoming just an ignition fuse to set off a much greater explosion. This greater explosion would result when hydrogen atoms were forced together ("fused") to form helium. This would be a "fusion bomb," a "hydrogen bomb," and "H bomb," whichever name you prefer.

The call went out, then, for liquid hydrogen—lots of it. But there were obstacles in the way—

Actually, hydrogen is a common substance. Two-thirds of all the atoms in petroleum, and in the ocean, are hydrogen. Three-fifths of the atoms in living tissue, including your own body, are hydrogen. Almost one atom out of thirty in the earth's solid crust is hydrogen.

However, hydrogen atoms are not found separately, but in combination with other atoms. To separate them from the others was at first a tedious and costly process. It was done by reacting certain metals with acid or by passing an electric

106

current through water. This sufficed for the small-scale 19th century uses of hydrogen.

Shortly after World War II, a group of oil companies and natural gas companies got together to try to set up a plan to make gasoline out of natural gas. They evolved a process of burning the natural gas and then quenching the flame at the proper point in order to make the burning incomplete and to produce carbon monoxide and hydrogen (rather than carbon dioxide and water). The carbon monoxide and hydrogen could then be recombined under the proper conditions and gasoline could be formed.

The process worked, but it proved uneconomic to produce gasoline in this manner in competition with the natural supplies of oil that became available after the war. However, the research had important ramifications. The new process proved to be far more efficient in the production of hydrogen than any of the older methods had been.

Consequently, when the call went out in the mid-20th century for a lot of hydrogen, more and more, the need could be met. Providing it in liquid form was another matter, however.

All through the 19th century chemists had tried to liquefy gases. Some, such as chlorine and sulfur dioxide, had yielded readily. A little cooling and those gases liquefied. In fact, a little pressure, even without cooling, was sufficient.

Other gases such as oxygen, nitrogen, and hydrogen did not liquefy despite considerable cooling and pressure. For a time, they were called "permanent gases." In 1869, however, chemists discovered that no amount of pressure would work unless the temperature was below a certain "critical point." For gases like oxygen, nitrogen, and hydrogen, this critical temperature was very low indeed.

Chemists therefore concentrated on lowering the temperature first and by the 1880s, oxygen and nitrogen were liquefied. Nitrogen was the more resistant of the two. Liquid nitrogen boils at a temperature of −320° F.—and even at that temperature, hydrogen remained a gas.

It wasn't until 1895, that the English chemist, James Dewar, managed to obtain liquid hydrogen. It boils at −423° F., a temperature that is only 38 Fahrenheit degrees above absolute zero—the very bottom of the temperature scale.

Liquid hydrogen could be formed, then; and if enough effort were put into the matter, large quantities could be formed. Yet for fifty years, it remained little more than an expensive laboratory curiosity.

The chief trouble was that this superfrigid liquid evaporated with superease. The most elaborate insulation did not help beyond a certain point, for liquid hydrogen generated its own heat.

The reason for this requires a little explanation. Under ordinary conditions, hydrogen exists as a collection of molecules, each molecule being made up of a pair of hydrogen atoms.

Each hydrogen atom consists chiefly of a central tiny particle called a "proton" which is constantly spinning. In some hydrogen molecules, the protons of the two hydrogen atoms spin in the same direction. That is "ortho-hydrogen." In other molecules, the protons spin in opposite directions. That is "para-hydrogen." In ordinary hydrogen gas, three-fourths of the molecules are ortho, the remaining are para.

Ortho-hydrogen contains more energy than para-hydrogen. When liquid hydrogen is formed, the ortho molecules slowly convert to the less energetic para. The extra energy of the ortho molecules is liberated as heat.

This slow conversion of ortho to para is constantly adding heat to the liquid hydrogen and evaporates it at the rate of one percent an hour, no matter how well it is insulated. What's more, if the container isn't properly vented, pressure can build up to explosive levels.

One way out seemed to be to get all the ortho changed over to the para to begin with. The hydrogen that remained would be pure para-hydrogen and, with proper insulation, that could be kept for long periods.

There are substances that will act as catalysts and hasten this conversion. As long ago as 1929, it was found that powdered charcoal would hasten the conversion, for instance. In 1952, under the pressure of sudden need, a preparation of iron oxide was discovered to convert large quantities of ortho-hydrogen to para-hydrogen in seconds.

This procedure was adapted to large-scale production and hydrogen can now be prepared in a form where, with proper insulation, one percent is lost through evaporation, not in one hour, but only after three days. The price has gone down to half a dollar a pound and liquid hydrogen plants are being built that will produce twenty tons a day and more. The call for liquid hydrogen was answered.

And the present needs for hydrogen will certainly match the supply, even though the needs continue to grow.

It would seem that one new use for hydrogen may lie in the production of electrical energy. Ordinarily, electricity is formed through a generator run by the heat energy of burning coal or oil (or, of course, by the energy of falling water). A great deal of energy is unavoidably lost in the passage from heat to electricity. If it were possible to combine fuel with oxygen in an electric cell set-up (a so-called "fuel cell") the process could be made much more efficient.

A number of fuels—including powdered carbon, carbon monoxide, and methane—have been tried in fuel cells. The practical difficulties involved in making such cells economic are great, but they are being overcome. The possibility that seems to hold most promise is the hydrogen-oxygen fuel cell. Such cells have been made to work, on a small scale at least, and the time may not be far off when hydrogen will make electricity cheaper and more available than ever before, in this particular way.

Liquid hydrogen, these post-war days, has a particularly exotic new use in "bubble chambers" employed to track down the strange and short-lived subatomic particles produced by the mighty atom-smashing machines of today. (These chambers were invented in 1952 by the American physicist, Donald W. Glaser.) One bubble chamber at the University of California is six feet long and contains 150 gallons of liquid hydrogen.

But fuel cells and bubble chambers can only use tiny quantities of hydrogen. The immediate use for all the liquid hydrogen even the modern vastly expanded program can supply will involve the rockets and missiles of today and tomorrow. In particular, liquid hydrogen will power the giant rockets that will point skyward to carry a man to the moon.

One post-war reason for preparing quantities of liquid hydrogen in a hurry has vanished. To be sure, the first experimental hydrogen bombs did use liquid hydrogen, but they were not practical. So much room and weight was taken up by insulation that the bomb was a monstrous and immovable creation.

The way out, apparently, was to use not hydrogen itself but a hydrogen compound with a light metal called lithium. This compound, lithium hydride, would explode just as hydrogen itself would, once it was ignited by a fission bomb. What's more, lithium hydride is a solid at ordinary temperatures and presents hydrogen in compact form that requires no pressures and no insulation. This made hydrogen bombs portable by aircraft and missile.

However, while we all hope that hydrogen bombs will never be used in anger, another aspect of the fusion process inspires no terror but holds out bright hope for mankind. If, somehow, hydrogen fusion can be brought under control and made to proceed slowly and steadily (instead of explosively), mankind's energy needs would be solved for the indefinite future.

What is needed is to raise the temperature of a quantity of hydrogen to the point where fusion will start and maintain itself; and to do this without the use of a fission bomb. It would help if we could find a way of making hydrogen undergo fusion at the lowest possible temperature.

To do so, it is necessary to make use of a rather rare kind of hydrogen. I said earlier that the hydrogen atom contains a central particle called a proton. One hydrogen atom out of every 7,000, however, carries along with the proton, a second particle called a "neutron." This "proton-neutron" hydrogen atom is twice as heavy as the ordinary "proton" atoms, so that it is called "heavy hydrogen." It is also called "deuterium" from a Greek word meaning "second" (because it contains a second particle along with the proton).

Deuterium was first discovered in 1932 by the American chemist, Harold C. Urey. Because of its double weight, deuterium was not very difficult to separate from ordinary hydrogen, but for ten years it remained only a laboratory curiosity. Then, during World War II, it was found that water containing deuterium ("heavy water") could be an important component of nuclear reactors.

As if that weren't enough, it was found, after the war, that deuterium would fuse much more easily than ordinary hydrogen. Consequently all efforts to tame the fusion reaction are concentrating on deuterium.

Even so, the temperature needed is in the hundreds of millions of degrees. At that temperature, deuterium atoms (and all other kinds, too) break up into a mixture of charged subatomic fragments called "plasma." Plasma is too hot to be contained by anything made out of matter, but since it is electrically charged, it can be confined by magnetic fields.

The problem is a tricky one, but each year we are raising the deuterium plasma to higher temperatures and keeping it confined for longer periods of time. Any year now (perhaps!) we will tame fusion.

And then, perhaps before the 20th century is over, a new kind of power plant will spring up here and there on the earth. Small containers of liquid deuterium will supply those power plants and fulfill the earlier functions of freight cars of coal and tankers of oil. It will be hydrogen, in one form or another, that will not only blast man's way to the stars, but will help power his conquest of hunger and misery on the face of the earth.

Chapter 11   Let There Be a New Light

In 1960, an American physicist, Theodore Harold Maiman, exposed a bar of synthetic ruby to strong light. After a while, the light which was absorbed by the bar was emitted again, but with a change. It appeared as a thin beam, deep red in color, flashing briefly out of one end of the bar.

That beam of light was of a variety never before seen by the eyes of man. As far as we know, it was a variety of light that had never existed on earth before, or in any part of the universe we can see. Maiman's bar of synthetic ruby was the first "laser," a device that we now look to as a possible death ray on the one hand, and as offering us peace-time miracles in surgery, photography, communications, space science, and half a dozen other fields on the other.

But what is it that makes the laser beam so different, unique? To the eye it merely seems a thin beam of colored light, and surely that has been seen before. What is there about it that the eye cannot see? To answer that question, let us first ask what ordinary light might be.

Suppose we picture light as a set of waves. We could ask, "Waves of what?" and get into trouble at once, but we don't have to ask that. It will suit us merely to imagine waves and that's all.

You mustn't think, though, that if you wanted to construct a wave picture of a beam of light, you must draw a wavy line that continues for the full length of a light beam. (The beams of light that reach us from the stars are many trillions of miles long, so "full length" can represent quite a figure.) Instead, we can picture the waves as being broken up into tiny lengths, each of which contains just a few up-and-downs, or "oscillations." We can refer to these tiny lenths of waves as "photons," an expression which comes from the Greek word for "light."

Photons are extremely small. A 40-watt bulb, which gives out a fairly dim light, emits about a quintillion (1,000,000,-000,000,000,000) photons each second.

Photons are not all alike by any means. The most important difference among them is that some contain more energy than others. Again, we can avoid asking embarrassing questions such as "What do you mean by energy?" and merely say that a more energetic photon can do things that less energetic ones cannot.

For instance, red light is made up of photons that are only half as energetic as those of violet light. When photons of red light strike ordinary photographic film, they lack the energy to cause the chemicals on that film to undergo changes. When the more energetic photons of violet light strike the film, the chemicals break down and the film is fogged.

That is why darkrooms in which ordinary film is developed may be lit by red light. The red light won't spoil the film.

Sunlight contains photons of a wide range of energies,

from red to violet and everything in between. It contains photons of all the energies that will affect the retina of our eye (a kind of living and very complicated photographic film) and beyond. It contains photons of infrared light, which don't register visibly on our eye and which are less energetic than any visible form of light. It also contains photons of ultraviolet light, which don't register visibly on our eye and which are more energetic than any form of visible light. (All the forms of light, visible and invisible, can be referred to as "electromagnetic radiation.")

Photons of ultraviolet light are so energetic that they can damage the retina, which is why it is dangerous to look directly at the sun. Photons of ultraviolet light are also energetic enough to bring about the changes in our skin that produce sunburn.

Photons of X rays and gamma rays, which are even more energetic than ultraviolet light, can smash their way right through our bodies and, if they make direct hits on certain molecules, can produce serious and even deadly chemical changes. That is why people working with radioactive substances or in modern atomic power plants, where such superenergetic photons may be encountered, must take extreme precautions against exposure.

Well, then, if we are going to picture photons as little bits of waves, we would want to be able to indicate the difference between one of high energy and one of low energy. This can be done by altering the length of each individual oscillation. You might draw a picture of a wave one inch long, and make the line of the wave curve so gently that you have only one oscillation in that inch. In the case of another picture of a one-inch wave, you may make ten oscillations.

The number of oscillations in a given length is called the "frequency" of the light. A wave with ten oscillations to the inch has a frequency ten times as great as that drawn with a single oscillation to that inch.

The greater the energy content of a photon, the higher its frequency. A photon of red light has about 35,000 oscillations per inch, while a photon of violet light has twice as many, about 70,000. (The difference in frequency of photons of visible light affects our eyes in such a way as to produce the sensation of different colors.)

Now let's see how photons are produced. To do that we must turn to the matter that makes up the universe.

Matter is made up of very tiny particles called atoms. These atoms, together with the still smaller particles that make them up, and the larger particles into which they can group, all contain energy. The energy content makes itself evident, very commonly, as motion. A high-energy particle moves or vibrates more rapidly than one of low energy.

Particles of matter don't just possess any amount of energy. They can possess only certain amounts; and each different type of particle can only possess certain amounts characteristic of themselves and of no others. Each particle can therefore be viewed as possessing certain characteristic "energy levels." The particle can be on one level or on another a bit higher up, but it can't ever be in between.

(The situation is similar to that involved in the coin system of money. Suppose a particular person had only nickels in his pocket. He could have 45¢ or 50¢ but he couldn't possibly have 47¢ in his pocket. If another person had only quarters, he might still have 50¢ in his pocket, but not 45¢.)

If a piece of wood is burning, the energy released by the combination of the particles of wood with air increases the energy content of the wood and air in the neighborhood of the fire. All the particles are kicked upward to a high energy level.

They don't stay at that high energy level, however. There is always a tendency for all particles to remain at the lowest possible energy level. Very soon, then, particles that have been raised to a high level fall back to a lower one. In doing so, they must give up the energy difference between the high level and the low one and this energy is given off in the form of a photon.

If all the particles near the burning wood were identical and if all moved up to the same high-energy level and then dropped back to the same low-energy level, then all the emitted photons would have the same energy content and would be of the same frequency.

114

This does not happen, however. There are all sorts of different particles involved, and they can move up to any of a large variety of energy levels. The result is that photons of a wide range of frequencies are given off, a few in the range of visible light, and we have a bonfire. Sunlight also is made up of a large variety of photon frequencies and so is almost any form of natural light. Until just a couple of decades ago, scientists took this incredible mish-mash of frequencies as an almost inevitable property of ordinary light.

But what if you begin with just one kind of particle and set up conditions that will make it possible for all the molecules to be at the same low-energy level? Suppose, further, these molecules are exposed to energy of just the right type to kick them up to the next higher energy level.

Every once in a while, a particle will, under these conditions, absorb enough energy to move up to the upper energy level and then fall back, releasing the energy as a photon of a particular frequency. There will always be some particles in the group that will have absorbed energy and that will be in the process of falling back. Photons, always of the same frequency, will always be streaming outward and such a system will therefore produce a beam of radiation of very constant frequency.

It was found, for instance, that ammonia gas could be made to emit a certain low-frequency type of radiation called "microwaves." This microwave radiation from ammonia had only about two oscillations per inch, as compared with the 35,000 oscillations per inch of red light.

These oscillations are very even and unvarying. They are far more constant than the oscillations of any man-made device such as a pendulum, and more constant, even, than the movements of heavenly bodies. In 1949, the American physicist, Harold Lyons, showed how to use these oscillations to control time-measuring devices and produce "atomic clocks" that were far more accurate than any others yet known. But there was more to such radiation than time-keeping.

The particles making up ammonia move from the lower energy level to an upper one when they happen to absorb a photon of just the right energy content. But what happens if an incoming photon strikes a particle that is already on the upper level? Does the particle rise to a still higher level? No!

In 1917, Albert Einstein showed from purely theoretical considerations that if a photon of the right size struck a particle in the upper level, it would not be absorbed. Instead, the particle it struck *would be knocked down to the lower level again.*

The struck molecule, moving down to the lower energy level, would produce a photon of just the same size as the photon that struck it. What's more, the photon that would be produced would be moving in the same direction as the original photon. You would begin with one photon coming in to strike the particle and would end with two photons of exactly the same frequency moving off exactly in step.

And what if each of these two photons happens to hit a particle in the upper level? Each particle would be knocked down to the lower level and produce two more photons, so that now you would have four photons, identical in frequency and all in step. Then if each of these strikes particles in the upper level—

But all this is not likely to happen under ordinary conditions because the particles remain at the upper level for only very brief periods of time. At any given instant, therefore, most of the ammonia particles are in the lower level and incoming photons are much more likely to hit low-level particles than high-level ones.

An American physicist, Charles Hard Townes, however, had thought of a way to separate the high-level particles from the low-level ones by means of an electrically charged device. In 1953, he managed to fill a small compartment with high-level ammonia particles only. If a photon of the right size happened to enter that compartment, it produced another photon. The two photons produced two more; the four photons produced four more; the eight photons produced eight more—

A single photon could trigger off a vast avalanche of identical photons in the barest fraction of a second. The device could thus be used as an amplifier. Suppose there was very feeble radiation from some point in the sky; radiation so feeble it could not be detected by our devices. If it hit the compartment of high-level ammonia, the resultant avalanche of photons could be easily detected and we could deduce the existence of the original photon (otherwise undetectable) that had sparked it off.

The original photon stimulated the emission of large quantities of microwave photons in order to produce amplification. Townes therefore referred to the device as something that produced "*m*icrowave *a*mplification by *s*timulated *e*mission of *r*adiation." The initials of the long words of that phrase were combined to form the word "maser."

The ammonia maser will only work for photons of one particular frequency, but there is no necessity of working only with ammonia. Solid substances were developed in which other combinations of energy levels were involved. In short order, masers involving a large variety of photon frequencies were developed.

At first, though, all masers could only work intermittently. The system would be pumped up to an upper level somehow and then an incoming photon would discharge it all the way. It would not work again until a new pumping had taken place.

A Dutch-American physicist, Nicolaas Bloembergen, got around this by devising a maser with a three-level system: a lower, a middle, and an upper. The system is pumped up by high-frequency photons capable of raising the atoms in the maser from the lower to the upper level. A second set of photons of smaller frequency is capable of knocking the system from the upper to the middle level, then from the middle to the lower. Both processes can work separately and constantly and the maser is pumped up as fast by one set of photons as it drops down to produce another set. For this reason, it can work continuously.

There is no reason why the radiation has to be of microwaves only. Why not make use of energy levels so far apart that much more energetic photons are produced; photons of sufficiently high frequency to register in the visible light region? A maser that produced visible light would be an "optical maser." Or else we could refer to it as bringing about "light amplification by stimulated emission of radiation" with the word "light" in place of "microwaves." With that change, the initials produce the word "laser."

Townes, in 1958, pointed out that a laser was completely possible in theory, and the first was constructed in 1960 by Maiman, as I said at the start of this chapter. Maiman's first laser was intermittent and had to be pumped up again after every brief discharge. Before 1960 was over, however, continuous lasers were prepared by an Iranian physicist, Ali Javan, working at Bell Laboratories.

Now we can see how the light of the laser beam is different from every other form of light we know.

First, the laser beam is very intense. In every ordinary light-producing process, a vast range of photon frequencies is brought forth. Only a small portion of them are usually in the visible light range. In the laser beam, *all* the energy released can be in the form of visible light, and the beam is an unusually concentrated form of light, therefore.

Second, the laser beam is very uniform. Ordinary light is made up of photons of a variety of frequencies, while the laser beam is entirely made up of identical photons. It is therefore all of one particularly tiny shade of one particular color. It is "monochromatic" light (an expression that comes from Greek words meaning "one color").

Third, the laser beam is very compact. The photons of ordinary light are moving every which way. It is difficult for that reason to keep a beam of ordinary light from spreading out. Laser beam photons, on the other hand, are all moving in the same direction. Ordinary light might be likened, therefore, to a vast mob with each member of it milling about in any direction he chose. The laser beam can, instead, be likened to a column of soldiers marching with absolute precision.

The natural tendency of the photons of the laser beam to move in the same direction is accentuated by the design of the tube that produces them. The ends are accurately machined to be absolutely flat and parallel. One is silvered so as to form a perfect mirror and the other is only lightly silvered. As the photons are produced by the laser action, there may be a number of avalanches produced in different directions. Most pass out the sides of the tube at once. Those avalanches, however, which happen to move along the length of the tube, strike first one silvered end, then the other, bouncing back and forth over and over, producing more photons constantly and building up a larger and larger avalanche. Any photon that, for any reason, isn't moving exactly parallel to the general line of the avalanche quickly hits one side of the tube or the other and moves out of the device.

Finally, when the avalanche grows large enough, it bursts through the lightly silvered end and you have the laser beam. The photons in that beam are so identical in frequency and direction that the oscillations in one photon seem to hook on to those in the photons before and behind and the result might almost be pictured as one long, long set of oscillations. The photons act as though they were stuck together, as though they "cohered" to one another. It is for this reason that the laser beam is spoken of as being made up of "coherent light."

A laser beam, made up of such coherent light, has virtually no tendency to spread apart. It sticks together and loses very little of its energy concentration as it travels through space. A beam of coherent laser light can be focused finely enough to heat a pot of coffee a thousand miles away. Laser beams even reached the moon in 1962, spreading out to a diameter of only two miles after having crossed nearly a quarter of a million miles of space.

The unique properties of laser light have made possible a large variety of interesting applications.

For instance, the narrowness of the beam of laser light means that a great deal of energy can be focussed into an exceedingly small area. In that area the temperature reaches extreme levels so rapidly that a necessary piece of work can be done before heat has a chance to radiate outward in quantities sufficient to do damage.

Thus, a flick of laser light into the eye can prevent some kinds of blindness by welding loosened retinas so rapidly that surrounding tissues remain unaffected by heat. In similar fashion, skin tumors can be destroyed without burning the skin.

A bit of metal can be vaporized and the vapor quickly analyzed by spectroscopic means. Holes can be punched in metals quickly and cleanly; even diamonds can be neatly reamed. Perhaps the laser beam will eventually help produce the extreme temperatures required to ignite a controlled hydrogen fusion reaction that will solve man's energy problems altogether (see Chapter 10).

Naturally—and sadly—the thought arises that what a small laser can do to a piece of thin metal, it might do to a human being. In 1965, lasers were developed which could be pumped to the higher level by the energy of chemical reactions. Can we not then imagine a pistol which does not use chemical energy to hurl a lead pellet but to emit a flash of a laser beam? It could strike a man with deadly effect, making no noise and leaving no tell-tale rifling marks. It would be a true "death ray" of the kind so often used in science fiction stories.

And if laser pistols, why not laser cannon? A gigantic laser could puncture, in a flash, the armor plating of a tank or ship. The light "missile" thus used would travel at 186,282 miles per second, and do so in a perfectly straight line, unaffected by wind, temperature, the rotation of the earth, the effect of its gravity, or any of the other items that make the precise aiming of material weapons so difficult.

The death ray has its limitations as a long-distance weapon. It can be weakened and absorbed or scattered by clouds, mist, smoke, and dust. Furthermore, its straight-line trajectory would not follow the curve of the earth so that it could not be aimed at anything beyond the horizon without bouncing it off a precisely positioned mirror.

If one looks through the crystal ball into the future, however, there arises the spectre of the use of such a death ray in space. In the vacuum beyond the atmosphere, there is no cloud, mist, or dust to interfere and no horizon to set bounds. Will mankind, a few generations hence, see space battles in which rocketships will flash laser beams at each other—with a momentary contact meaning a puncture?

Such laser beams would require vast amounts of energy but lasers are now being developed which derive their energy from sunlight. Out in space, the lasers could be powered without limit by the ever-present, never-shrouded sun.

But let us hope society advances to the point where such

weapons, large or small, are never needed or used. There are sufficient peacetime uses to keep the laser beam busy. They can be applied to the communications industry, for instance, an industry now heavily dependent on the very low-frequency photons of microwaves and radio waves.

These low-frequency photons can be "modulated," that is, the photon stream can be made to vary in a regular fashion, so as to produce mechanical vibrations in a diaphragm that will in turn produce sound waves in the air. Or they will produce variations in an electric current, which will, in turn, produce light of varying intensity. It is this which gives us the sound of radio and the sight and sound of television.

To prevent interference of one message with another, different messages must be carried on "carrier beam" photons of markedly different frequencies. In the low-frequency region there aren't very many different frequencies available, and the number of radio stations or television channels that can be broadcast is therefore quite limited.

If light photons were used as carrier waves, their far higher frequencies would allow room for many more message ranges. (We can see why this is so if we consider the range of numbers from 1 to 10 as representing radio waves, and those from 1,000,000,000 to 10,000,000,000 as representing light waves. In both ranges, the last number is ten times as large as the first; yet in the range from 1 to 10 there are only ten integers, while the range from 1,000,000,000 to 10,000,000,000 includes nine billion and one.)

For radiation to act as a carrier wave, it must be very even in frequency and direction. This was possible for the gently oscillating radio waves but not for the very high-frequency light waves—until the laser was invented. Of course, it is not easy to modulate the light waves of a laser beam, but the problem is being solved. In 1965, New York's seven television channels were transmitted across the width of a room on a single pencil-thin laser beam and each channel could be sorted out from the others thereafter.

Will the time come when laser beams, reflected and amplified by communications satellites, will serve the world? If that were possible, there would be room for all the different radio stations and television channels in the world, and as many more as men might want to build.

Possible atmospheric interference to such a system would not apply to space. Space ships and space stations could communicate with each other, and also with stations on the surface of an airless world like the moon, by messages carried on laser beams.

Indeed, the information sent would be not only a matter of words alone. The absolutely straight line of the laser beam would serve to locate the precise position of one ship or station with respect to another at some precise instant of time. Furthermore, the laser beam would be reflected off the ship or other object being observed and the reflected light would change in frequency very slightly depending on whether the reflecting object were moving away or moving toward the observer, and how quickly. The beam would also be affected by whether the observed object were rotating, in which direction, and how quickly.

The same could be done with ordinary light, to be sure, if ordinary light could be packed into a tight enough beam of sufficient energy to reach across space and bounce back without too much loss. However, ordinary light contains photons of so vastly many frequencies that slight changes in those frequencies could not be detected, as they could be in the photons of the laser beam. (If every man in a large active crowd quietly took a tiny side step to the left while he was milling about, could that be detected? If a line of soldiers, marching in precise formation, all took that same side step, could that be detected? No, in the first case; yes, in the second; I think you will agree.)

It may well be that when the space age reaches maturity, a truly enormous load of communication and information will be carried by the laser beams interlacing space between the various human outposts. It is very likely that it will then be maintained that space exploration could never have progressed beyond the most primitive hit-and-miss stage without the laser.

Getting back to earth, we find that a recent application of the laser involves photography. In ordinary photography, light is recorded on plates or films through the effect of that light upon chemicals. The more intense the light the greater the effect. The chemicals, therefore, record brightness and produce a replica of the bright-and-dark pattern of the light emitted by an object or by light reflected from it. That replica is the photograph.

Suppose though that a laser beam is sent against a mirror which reflects it back without distortion onto a photographic plate. Another laser beam is simultaneously reflected from some ordinary object which reflects it, with distortion, to the photographic plate. (The distortion arises from the fact that the ordinary object has an uneven surface, so that some parts of the laser beam are absorbed, some are not; some are reflected in one direction, some in another.)

At the photographic plate, the two beams, one undistorted and one distorted, meet. The total intensity of light at each spot is recorded as in ordinary photography. In addition, however, the waves of the two beams crisscross in a variety of ways that depend on the exact details of distortion in the beam reflected from the ordinary object. Such crisscrossing is called "interference." The plate records not only the intensity of the light but also the interference pattern.

Physicists knew this was possible many years ago, but ordinary light won't do. All the different waves in ordinary light, moving with different frequencies in different directions, would produce such an interference mish-mash that no useful information could be extracted from it.

With a laser beam, however, a neat interference pattern, depending only on the nature of the reflecting object and nothing else, can be produced. The plate has *all* the information, intensity and interference, and the process is called "holography" ("holo-" meaning "whole").

The plate, or "hologram," carrying all this information shows nothing to the eye except, sometimes, a pattern of circles that arises from dust specks. The interference pattern is microscopic.

If a laser beam is sent through the hologram, an image of the original reflecting object is created; a partially three-dimensional image that can be photographed from different angles. This was first carried through in 1964, and by 1966 it was no longer necessary to use a laser beam to create the image; ordinary light would do, so that the process became cheaper and more practical. (However, a laser beam remains necessary to form the hologram in the first place.)

A hologram can be made of a very quickly moving object or just a briefly existing one, and it then provides a permanent image that can be studied with much more detail than a mere photograph can. Holography gives much finer detail, too, and scientists are looking forward to a time when holographic microscopy will introduce us to the world of the tiny with a new clarity.

And, a little further out into the wide blue, comes the thought that perhaps holography can be perfected to the point where a complete 3-D image can be formed and the process built into a television set.

Will the day come when we will no longer have to content ourselves with a two-dimensional television screen marked out in coarse lines of light and dark, but will see a true and perfect three-dimensional color representation?

In some future Miss America contests will the girls parade through a cube in our living rooms in three dimensions? To be sure, the girls will be images only, nothing more than focussed, impalpable rays of light. They won't be *real* girls. Even so, how nice it would be!

Chapter 12   The Ocean Mine

Our mines are being used up. Our population is increasing by leaps and our industrial production by bigger leaps— and the mineral resources of the United States suffer the consequence. Our best copper mines are gone. Our best iron mines are going. We have to learn to deal with poorer ores and make them do.

But the situation isn't entirely black. For some types of mineral resources, the largest and richest mine that ever existed lies at our doorstep and is still hardly touched. That mine is the ocean.

In area, the ocean covers more than 140,000,000 square miles, about seven-tenths of the earth's surface. Its average

This article appeared in *Science World*, March 19, 1957.

depth is two and a third miles, so that the total amount of sea water on the earth is equal to 330,000,000 cubic miles.

What makes the ocean a mine is the fact that those hundreds of millions of cubic miles do not consist only of water. If you've ever done any ocean bathing, you know that the ocean isn't just water. "Just water" doesn't taste the way sea water does.

The fact is that about 3.25 percent of the ocean is solid matter held in solution by the remaining 96.75 percent which *is* water. This is really a great deal of solid matter, and we don't have to consider the whole ocean to prove that. A smaller amount of sea water will do—say, a swimming pool full.

Imagine a swimming pool fifty feet long, thirty feet wide, and averaging six feet deep. If it were filled with sea water, it would contain 285 tons of liquid; of this, nine and a quarter tons would be dissolved solids. To put it another way, if the water in the swimming pool were evaporated, nine and one-quarter tons of solid matter would be left at the bottom. That is a respectable amount for a swimming pool.

As you can tell by the taste of sea water, most of the solid matter is ordinary salt—sodium chloride. Nearly seven and three-quarters tons of the solid matter in our pool would be made up of sodium chloride, and another three-quarters of a ton is made up of chlorine atoms in combination with metals other than sodium.

Leaving this to one side, there is still about three-quarters of a ton of material in our dried up swimming pool of sea water that is neither sodium nor chlorine. Properly treated, that three-quarters of a ton of miscellaneous material could yield: 750 pounds of magnesium, 500 pounds of sulfur, 230 pounds of calcium, 220 pounds of potassium, 37 pounds of bromine, and about 28 pounds of other things, including every element in existence—copper, silver, gold, uranium, and even radium.

Of course, there is a catch. To get minerals out of the ocean we must concentrate thinly spread atoms. This takes energy, one way or another. The less concentrated the min-

eral we're after, the greater the energy required to extract it. There's no way of getting around that.

Fortunately, the sun itself has done the job for us in
many cases. Every once in a while during the course of the
ages, a shallow arm of the ocean has been pinched off by
rising land. If the climate is such that the inland sea thus
formed evaporates at a rate faster than fresh water can be
brought to it by rivers, it gradually shrinks. The salts it
contains become more and more concentrated, and the sea
may eventually dry up completely, leaving its solids behind.

Salt mines are the residues of dried-up portions of the
ocean. And we all know how important salt has become. It's
not just something to be put on food (though that is essential). It has hundreds of important industrial uses, and it
is the main source of such important chemicals as chlorine
gas, hydrochloric acid, sodium hydroxide, sodium carbonate,
and many others, each with additional hundreds of uses.

If an inland sea dries out slowly, salts are deposited in
layers. The reason for this is that sodium chloride is one of
the less soluble salts in the ocean. Also, it is present in the
greatest concentration. As the inland sea dries, therefore,
sodium chloride begins to precipitate even while there is
still more than enough water to keep the other salts in
solution. Then, in the last stages of evaporation, the other
salts precipitate on top of the sodium chloride. The sun has
not only taken out the solids for us, but separated them into
layers, too.

The salt deposits near Stassfurt, Germany, are well-known
examples of this layering process. They were the best source
of potassium salts in the world and, because of them, potassium salts were much cheaper in Germany than anywhere
else. In northern Chile, there are dried-up salt beds which
are rich sources of sodium nitrate and potassium nitrate.
Before World War I, these salt beds were the main source
of nitrates for the manufacture of fertilizers and explosives.

Then there are inland seas that are in the process of drying
up. Salt deposits have formed all along the edges of these
seas, and what water is left in them is thick with dissolved
materials. The best-known examples of such drying seas are
the Dead Sea on the Israel-Jordan border and the Great

Salt Lake in Utah. The minerals of the Dead Sea are an important Israeli resource.

Then, too, there are numerous salt marshes and underground regions of water with a high salt content. These are called "brines" and are sometimes associated with oil wells. Iodine can be obtained in commercial quantities from such brines.

But what about the chances of extracting minerals from the ocean directly? Would it be possible for scientists to develop an artificial "drying up" process?

It's possible. At least two different elements are now produced by man from sea water in all the quantity we need.

One of these is magnesium. Its atoms—after those of sodium and chlorine—are the most common in the solid matter of the ocean. To extract magnesium, sea water is pumped into large tanks and calcium oxide (lime) is added. (The calcium oxide comes from the sea, too, since it is produced by roasting oyster shells.) The calcium oxide reacts with water and with magnesium ions in solution. Magnesium hydroxide is formed and this precipitates as a solid.

The magnesium hydroxide is filtered off and converted to magnesium chloride by reaction with hydrochloric acid. The magnesium chloride is then passed through filters and driers and finally converted by means of an electric current to magnesium metal and chlorine gas. (The chlorine is converted back to hydrochloric acid for use on the next batch of magnesium hydroxide, so it's not wasted.)

The other element that is commercially mined from the sea is bromine. This is more difficult to mine than magnesium. There is only a twentieth as much bromine in the ocean as there is magnesium. Nevertheless, the problem of collecting it has been solved by scientists.

To do the job, a large quantity of sea water is made acid, and chlorine gas is added to it. (Both hydrochloric acid and chlorine gas are made from salt, which comes from the sea.) The chlorine reacts with bromide ions in the sea water and must be blown out with air. That is, air is bubbled through the sea water, and, as it comes out, it carries some bromine vapor with it. The air is then passed through tubes packed with sodium carbonate: the bromine gas is absorbed, sodium bromide and sodium bromate being formed. Now the bromine is concentrated into a relatively small volume. It can be, and is, easily separated.

There is still a third element obtained from the sea, but more indirectly. That element is iodine. The amount of iodine in the ocean is less than a thousandth the amount of bromine. Our swimming pool of sea water that would yield up thirty-seven pounds of bromine would supply us with not quite half an ounce of iodine. That's too little for our industrial techniques to concentrate profitably.

Too little for *man's* techniques, that is. But in the water there are living organisms, such as seaweed, that need iodine for their own life processes. Patiently, these organisms collect the iodine atoms out of the sea water that streams past and through their bodies. It is then only necessary for man to farm the sea for seaweed. The weed is burned in shallow pits, and the ash, called "kelp," turns out to be one per cent or more iodine. The ash has two hundred thousand times as great a concentration of iodine as the sea itself and is a commercial source for that element. (In fact, iodine was first discovered in seaweed ashes back in 1811.)

The sea is an inexhaustible source of such elements. Not only is the quantity so great as to satisfy all man's needs for a long time, but even when extracted, the ocean has not lost it permanently. Compounds of all elements are constantly washing back into the ocean thanks to rain and rivers and what we take out will return.

It looks as though other elements than the three mentioned may eventually come from the sea. Nor will they have to be concentrated. It turns out that large areas of the ocean floor are strewn with metallic nuggets rich in manganese and containing reasonable quantities of such valuable metals as cobalt, nickel, and copper. Dredging operations many miles off-shore may become a common sight in the future.

To give an idea, incidentally, of how vast a mine the ocean really is, let's do some more calculation. One cubic mile of ocean is enough to fill more than sixteen million swimming pools of the size I mentioned earlier in the chapter and there are, as I have said, 330,000,000 cubic miles of ocean all told. It is not surprising (or should not be), then, that the ocean contains five hundred quadrillion (500,000,-000,000,000,000) tons of solids. This includes:

2,000,000,000,000,000 tons of magnesium,
100,000,000,000,000 tons of bromine, and
75,000,000,000 tons of iodine,

plenty of each to last mankind for a long time.

The ocean contains surprising amounts of a few other metals in solution (in addition to what may be contained in the nodules on the ocean floor). It contains, for instance:

15,000,000,000 tons of aluminum
4,500,000,000 tons of copper
4,500,000,000 tons of uranium
1,000,000,000 tons of thorium
450,000,000 tons of silver
45,000,000 tons of mercury
6,000,000 tons of gold, and
45 tons of radium.

These quantities, though large, are so thinly spread throughout the ocean that we know of no way yet of collecting them profitably.

Chapter 13   Our Evolving Atmosphere

The 1960s have brought us new knowledge concerning the atmospheres of our planetary neighbors. Observations from balloons floating high above our own interfering air have produced important evidence leading us to the belief that the clouds on Venus are composed of ice particles. Mariner IV, as it skimmed past Mars in 1965, told us that its atmosphere was considerably thinner than had been thought.

But all our latest observations merely served to confirm what astronomers have been suspecting for quite a while now. The atmosphere of our own earth is unique; it is not like any other atmosphere within range of our instruments.

The planetary atmospheres we know fall into four classes:

First, a planet or other cold body may have no atmosphere at all, or one so thin that it can scarcely be distinguished from the vacuum of space.

Second, an atmosphere may be rich in hydrogen and related compounds which encourage the type of chemical reactions called reductions. This would be a "reducing atmosphere."

Third, an atmosphere rich in free oxygen would be an "oxidizing atmosphere."

Fourth, an atmosphere may contain neither hydrogen nor oxygen but only gases that bring about neither oxidation nor reduction. This would be a "neutral atmosphere."

The planets of our solar system (excluding Pluto, concerning which we have no atmospheric information) fall into the following classes:

(1) little or no atmosphere: Mercury

(2) reducing atmosphere: Jupiter, Saturn, Uranus, and Neptune.

(3) oxidizing atmosphere: Earth

(4) neutral atmosphere: Venus and Mars.

Of the 31 satellites in the solar system, Titan (the largest satellite of Saturn) is the only one known to have an atmosphere—a reducing atmosphere. All the rest, including our moon, have, as far as we know, little or no atmosphere.

In short, nowhere in the solar system but on our own earth can we find an oxidizing atmosphere. Nowhere else is there free oxygen.

Why?

Suppose we begin with the cloud of dust and gas out of which it is thought the solar system developed. Astronomers think that about 90 percent of it was hydrogen and another 9 percent helium. The remaining 1 percent consisted of oxygen, neon, nitrogen, carbon, silicon, magnesium, iron, sulfur, and argon, probably in that order of decreasing quantity, with a scattering of the remaining, even less common, elements.

Carbon, silicon, magnesium, iron, and sulfur are solids at

ordinary temperatures and form solid compounds among themselves (carbides, silicides, and sulfides). As the cloud swirls into eddies, atoms and molecules of these elements and compounds tend to stick together. First pebbles, then rocks, then larger bodies called "planetesimals" formed. These eventually form the solid core of a planet. On earth, a large excess of metallic iron eventually settled to the very center while rocky substances formed a thick outer shell.

Hydrogen, present in overwhelming quantity, would combine with almost everything in sight. It would tie up most of the oxygen as water molecules (H_2O), the nitrogen as ammonia molecules (NH_3), much of the carbon as methane molecules (CH_4) and some of the sulfur as hydrogen sulfide (H_2S). Hydrogen could not combine with helium, neon, or argon, however, for these last three are "noble gases" that form no compounds as far as we know.

These substances—hydrogen, helium, neon, argon, water, ammonia, methane, and hydrogen sulfide—are all low-melting and, at ordinary temperatures, are gases or (in the case of water) an easily evaporating liquid.

At the low temperatures at which the planets first formed, some of these substances, particularly water and ammonia, may have been solid and may have collected with the metal and rock as the planetesimals formed. Even those substances which remained gaseous may have been tapped in considerable quantity within the loose structure of the gathering solid.

But then, at the center of the solar system, the huge interior mass of the cloud had condensed to the point where internal temperatures caused it to break into nuclear flame. The sun was born.

Increasing heat from the sun vaporized the low-boiling substances, and gases bubbled up from the fragments of the forming planet. Such gases don't cling to the structure of the planet by chemical bonds; they must be held by gravity alone. If gas molecules move slowly, they tend to be held by moderate gravitational forces; if they move rapidly, they tend to be lost.

The hotter a gas, the more rapidly its molecules move and the more easily it is lost. The groups of planetesimals closest to the sun felt the heat of the solar furnace most intensely and their atmospheres began to drift away. The gases were swept outward by the gathering solar wind (consisting of particles blasting outward from the sun) and carried into the outer, cooler regions of the solar system.

As the outer planets formed they grew huge with gas; not only with their own gas, whose slow-moving cool molecules they could retain, but also with the gas reaching them from the inner solar system. It is for that reason, that Jupiter, Saturn, Uranus, and Neptune (particularly Jupiter which had "first pick" at the outward-flowing gas) are so much larger than the inner planets. Their substance consists largely of the hydrogen and hydrogen compounds that made up so much of the original cloud. They have thick reducing atmospheres of hydrogen, helium, ammonia, and methane.

The inner groups of planetesimals, however, had lost their original "primary atmosphere" completely. Free hydrogen and the inert gases were gone forever. Some of the molecules of water, ammonia, methane, and hydrogen sulfide must have managed to hang on to the solid core of the planetesimals by forming loose chemical combinations with their structure.

But the inner group of planetesimals were still coalescing into planets, and their gravitational fields were growing steadily more intense. Their interiors were heating up, and gaseous molecules were slowly forced out of combination and sent fizzing up out of the interior, pushed out by pressure or belched out through volcanic action. Mercury never grew large enough to hold much of these gases against the heating action of the nearby sun, thanks to its small size and weak gravitational field. It, therefore, has little atmosphere at the present day.

The other interior planets, Venus, earth, and Mars, grew larger and remained cooler than Mercury and managed to hold some of the gases. Relatively small Mars could hang on to only a thin layer, but earth and Venus did better. Molecules of ammonia, methane, and hydrogen sulfide coated Venus, earth, and Mars with a thin "secondary atmosphere" that was reducing in nature.

Water was also forced out of the planetary interior. Some of it remained in the atmosphere as vapor but most of it condensed into liquid. On earth, large oceans slowly formed;

but there was apparently less water retained on the slightly smaller and warmer Venus, and still less on the considerably smaller, if cooler, Mars.

It was under a reducing atmosphere, then, that life developed. Indeed, life *required* such an atmosphere to develop (see Chapter 9). In order for life to form, complex molecules, made up chiefly of carbon and hydrogen atoms, had first to build up. These could not have formed spontaneously in the presence of the free oxygen with which our atmosphere is loaded today.

Furthermore, these complex molecules could only form at the expense of the sun's energetic ultraviolet radiation, bathing the early atmosphere and ocean. Had the atmosphere contained oxygen, the necessary ultraviolet would have been shielded off and its energy would not have been available.

Of course, the same ultraviolet that supplies the energy for the formation of complex carbon-hydrogen molecules in the first place would tend to break down those that become particularly complex. Eventually, then, simple life forms made up of very complex molecules would fill the seas a few dozen feet below the surface, hovering at a level to which the ultraviolet cannot penetrate. In the uppermost level, the moderately complex molecules would form and these, percolating downward, would serve as food for the life forms.

But even while life was developing, the atmosphere would be continuing to evolve. The ultraviolet radiation, streaking through the atmosphere, would strike water molecules and break them apart into free hydrogen and free oxygen ("photodissociation").

The smaller the mass of a gas molecule, the more rapidly it moves at any given temperature and the more easily it escapes from a gravitational field. Hydrogen atoms are the lightest known and they move too rapidly to be held long by earth's gravitational field. The hydrogen atoms liberated by the breakup of the water molecule slowly leak away, therefore, into interplanetary space and are gone.

The atoms of free oxygen, massive enough to be retained by the earth's gravitational field, combine to form oxygen molecules (each made up of two oxygen atoms) and these combine further with other substances. They combine with

133

rocks of the soil to form oxidized minerals—chiefly silicates. They also combine with the ammonia, methane, and hydrogen sulfide molecules in the atmosphere, forming nitrogen and water in the first case, carbon dioxide and water in the second, and sulfur and water in the third.

The water that is formed in these reactions is photodissociated in its turn and that keeps the process going. Sulfur joins the solid core of the planet to form sulfides, or, after combining further with oxygen, to form sulfates. The ammonia and methane of the atmosphere gradually change to nitrogen and carbon dioxide, then, at the expense of a steadily decreasing water supply. The reducing atmosphere changes to a neutral atmosphere.

This happened on Mars, whose thin atmosphere is almost entirely carbon dioxide now, and whose water supply has decreased to the point where it is barely sufficient to form thin icecaps at the poles.

On Venus, the atmosphere is now probably made up of nitrogen and carbon dioxide. While Venus retains considerable water even now, some estimates place its present water supply at only 1/10,000 that of the earth's oceans.

Since Venus always had a much thicker atmosphere than Mars did, Venus now contains much more carbon dioxide in its atmosphere and that is crucial.

Carbon dioxide does not absorb visible light to any great extent, but it absorbs infrared radiation strongly. Sunlight passes right through an atmosphere containing carbon dioxide, hitting soil and ocean, and being absorbed as heat. The heated surface radiates away some of the heat as infrared, but this radiation is absorbed and retained by the carbon dioxide of the atmosphere, which heats up in consequence.

A planet with an atmosphere poor in carbon dioxide and other absorbing gases would allow infrared to escape out into space and would remain cool, while another planet with an atmosphere rich in carbon dioxide would retain infrared and grow hot—even though both planets were equal distances from the sun. This action of carbon dioxide is called the "greenhouse effect" because the glass in a greenhouse also acts to transmit light and retain infrared, keeping the interior warm and damp even in the winter.

As Venus's reducing atmosphere became neutral and as

OUR EVOLVING ATMOSPHERE

more and more carbon dioxide was formed, that atmosphere
grew hotter and hotter. Eventually, the temperature of both
atmosphere and planet rose to the point where what water
was left boiled away to form the clouds that now cover
Venus eternally. Water vapor also absorbs infrared radiation and the presence of the cloud layer on Venus further
intensifies the greenhouse effect.

It might seem that this process could continue further,
assuming that there was enough water to begin with. Oxygen could continue to pour into the atmosphere, and when
all the ammonia and methane had been converted to nitrogen and carbon dioxide, and all the surface rocks had been
converted to silicates, additional quantities of oxygen would
begin to accumulate as such in the atmosphere. This, however, is not so.

As soon as free oxygen begins to enter the atmosphere, it
begins to absorb ultraviolet radiation. In the process, the
two-atom molecules of ordinary oxygen are converted to the
more energetic three-atom molecules of ozone.

A layer of ozone forms in the upper atmosphere and ultraviolet is absorbed by it. The ultraviolet penetrating the layer
and reaching the lower atmosphere, where water vapor exists, decreases in amount steadily as the ozone concentration
builds up, and eventually photodissociation slows to a stop.
Photodissociation is a "self-limiting process." It can convert
a reducing atmosphere into a neutral atmosphere and seems
to have done so on Mars and Venus, but it cannot go on to
form an oxidizing atmosphere.

How then did earth's oxidizing atmosphere originate?

To begin with, there must wave been photodissociation
on earth as on Venus—though probably at a slower rate,
since earth is farther from the sun than Venus is and receives ultraviolet in smaller doses. Even so, earth's water
supply began to decrease and its atmosphere began to grow
neutral, and in the end perhaps half of the planet's total
supply of water was lost. Fortunately, earth could afford
the loss and enough water was retained to form our copious
ocean of today.

Yet somehow the process did not end as it did on Venus.
A new factor entered and that factor seems to have been a

135

chemical development that appeared in some of the life forms in the oceans of the primitive earth. Without that development any simple life forms that may have developed on Mars must have been able to just barely hang on as the planet slowly dried up. And any simple life forms that may have developed on Venus would have been even less fortunate for they must have been killed outright as the planet slowly heated up and boiled.

The life forms on earth would have had to face similar ends, if nothing new had come up. They were tiny creatures at the time, made up of single cells not much larger or more complex than present-day bacteria. They drifted idly below the upper layers of the ocean, living on the drizzle of complex food molecules from above. They lived only at the pace made possible by the slow production of food by the sun's ultraviolet light.

Then came the development of a molecule called "chlorophyll." This molecule is built around a complex but stable ring of atoms that could be built up out of simpler molecules by ultraviolet light. Occasionally, frills would be added in the form of short "side chains" of atoms attached here and there on the ring. One particular combination of side chains produced chlorophyll. This was capable of absorbing visible light, in the red range particularly. Green was reflected, and chlorophyll is therefore a bright green in color. When chlorophyll absorbed visible light it became charged with energy and this energy could bring about certain chemical changes.

Once cells incorporated chlorophyll into their structure, they had an important tool for producing changes they couldn't manage before. Indeed, they could use the energy of visible light, after it had been stored in the chlorophyll molecule to produce a series of changes that ended in the production of the complex food molecules, upon which the cell could feed without having to wait for the general drizzle. This process is known as "photosynthesis" ("build-up by light").

One of the consequences of photosynthesis is that the energy of *visible* light is made to bring about the breakdown of water molecules into hydrogen and oxygen. Visible light, unlike the more energetic ultraviolet, would not do this in the absence of chlorophyll.

Water breaks down far more rapidly through the effect of concentrations of chlorophyll systems within cells than through the hit-and-miss action of ultraviolet. Cells that made use of chlorophyll obtained more food and could multiply more rapidly than could cells that didn't make use of chlorophyll. Gradually, over many years, the use of chlorophyll became nearly universal and photosynthesis became the prevailing mode of life. Since the chlorophyll-containing cell was green in color, the world of life slowly turned green. And our planet has remained the good green earth to this day.

It might seem that photosynthesis only speeded the break-up of water and merely hastened the conversion of a reducing atmosphere to a neutral one. Actually, it does more!

Evolution could now *pass* the stage of the neutral atmosphere. Once the neutral atmosphere had formed completely and additional oxygen began to exist free, a protective umbrella of oxygen did indeed form in the upper atmosphere and eventually formed an ozone layer. Ultraviolet light began to be shielded off and photodissociation dropped off steadily. But *visible light* could pass through the ozone layer so that photosynthesis could still continue. Unlike photodissociation, photosynthesis is not a self-limiting process in this respect. More and more oxygen continued to pour into the air, and earth's atmosphere passed beyond the neutral stage and began to become oxidizing.

But even so, would not the high concentration of carbon dioxide also present in the atmosphere trap the sun's heat and boil the earth's oceans as those of Venus had been boiled?

Fortunately, the breakup of water molecules is not all that is involved in photosynthesis. The hydrogen molecules that are formed in the course of the process do not enter the atmosphere and gradually leak out into space. Instead, the hydrogen undergoes a series of chemical reactions which end in its combination with carbon dioxide to form starch and other components of plant cells.

Thus, while photosynthesis pours oxygen into the atmosphere, it does not allow hydrogen to escape but uses it to remove carbon dioxide from the atmosphere. In the end, earth's atmosphere is composed almost entirely of nitrogen and oxygen.

Exactly when this change began to take place is not known. The best guess, based on the chemistry of ancient rocks, is that free oxygen began to enter the atmosphere between one and two billion years ago, when life on earth had already existed for one or two billion years.

By some 600,000,000 years ago, the amount of oxygen in the atmosphere had grown to be at least one-tenth what it is now. This brought about a biological revolution and ushered in what geologists call the Cambrian period.

During the pre-Cambrian stretch of time, when little or no oxygen had been present in the atmosphere, life forms had obtained energy out of complex organic molecules by breaking them down to simpler structures without any particularly radical change in the nature of their chemical structure. This process is "fermentation."

With a reasonable supply of oxygen in the atmosphere, however, some twenty times as much energy could be obtained by life forms that developed systems for combining food-stuffs with oxygen.

With huge quantities of new energy available, life flourished and proliferated. During the hundred-million-year Cambrian period, life forms grew complex and developed into myriad forms.

Cells clung together to form multicellular organisms. Different groups of cells within these organisms could specialize. Some developed methods of rapid contraction and other methods of conducting electrical impulses, so that muscles and nerves could form. Shells and other stiffening agents were grown to keep the large masses of cells from collapsing and to protect the organism from its enemies. There seemed suddenly no end to the ingenious development life forms could undergo once plenty of energy was at hand.

The shells and other hard structures remained behind after the organism had died. They assumed a stony structure over the eons and the rocks of the Cambrian period are rich in such remnants, called "fossils," whereas rocks dating back to earlier times are fossil-free.

By 400,000,000 years ago, the content of oxygen in the atmosphere probably reached its present level. The ozone umbrella was tight, and the amount of ultraviolet reaching earth's surface was sufficiently low to allow life forms to withstand direct sunlight for reasonable periods of time.

For the first time, life forms could venture out onto dry land and the continents could be colonized.

But the evolution of the atmosphere did not come to an end with the formation of what we now have. There have been fluctuations in this component or that, and the fluctuations with the most important consequences have been those involving carbon dioxide.

At present, only 0.03 percent of the atmosphere is carbon dioxide, but its importance is out of all proportion to its quantity, not only because it is the ultimate food supply of plant life (and therefore of animal life as well) but because of its greenhouse effect. Even small changes in carbon dioxide concentration can have a powerful effect on the earth's temperature.

There are periods in history when vast surges of volcanic action all over the earth pour unusual quantities of carbon dioxide into the atmosphere and raise its concentration somewhat. The atmosphere begins to retain more heat and the earth grows warm. With the warmth and with increased quantities of carbon dioxide, plant life flourishes and vast forests cover the land. It may have been after periods like that that the planet's vast coal fields and oil pools have formed, out of those forests.

At other times, mountain-forming periods heave huge masses of rocks to the surface. These fresh rocks, not previously exposed to air, combine with carbon dioxide to form carbonates. The carbon dioxide content of the air drops to abnormally low values, the greenhouse effect is diminished and the earth cools off. If it cools off more than a certain amount, ice ages may follow. We are at the end of a longish period of mountain-building and ice ages now.

But mankind is now ready to introduce a new factor, never before present on earth—his own technology.

Man is digging out of the ground the coal and oil that

were slowly laid down over many millions of years, and is burning it all in the course of a century or two. He is forming, once again, the carbon dioxide which plants had long ago consumed to make the tissues that eventually changed into coal and oil.

Six billion tons of coal, oil, and gas are being burned each year and the quantity of carbon dioxide in the atmosphere is slowly creeping upward (even though much of it is dissolved in the ocean and consumed by plants). It is estimated that at the rate we are going, the quantity of carbon dioxide in the air will be 25 percent higher in the year 2000 than it is now. By 2300, the quantity may have doubled.

The presence of 0.06 percent of carbon dioxide in the air will in no way poison us, but what about the greenhouse effect? Slowly, the earth's average temperature will creep upward. Indeed, it has risen a bit in the first half of the 20th century, possibly because of the added carbon dioxide in the air.

If the earth grows slightly warmer, we may expect the polar icecaps to begin to melt and the added water will cause the ocean level to rise. Even if we allow for the fact that the added weight of water would tend to depress the sea-bottoms, we can still expect sea level to be 200 feet higher than today once all the icecaps melt.

All the coastal regions of the continents, just where one finds the greatest load of human population, would be water-covered. It has been calculated, however, that even under the most extreme conditions, it would take 400 years for the icecaps to melt completely and mankind might have time to plan some action. A wholesale switching from coal- and oil-burning to nuclear fuels might help. Devices for removing carbon dioxide from the atmosphere in large quantities might keep the earth cool; so might a scheme for spreading the oceans with substances designed to reflect sunlight more efficiently.

As a last resort, we might transfer populations, for certain regions, nearer the poles; and deserts, which now support little life, may become capable of maintaining large populations.

Earth's atmosphere, which has presented life forms with crises in the past, may be on the verge of presenting us with an enormous one in the near future.

Chapter 14   The Atmosphere of the Moon

Now that satellites are approaching the moon, circling the moon, landing on the moon, and now that we are getting ready to send men themselves to the moon, any bit of information about the moon is helpful. What about the moon's atmosphere, for instance?

But the moon has no atmosphere, you say.

Certainly, it doesn't have any atmosphere in the sense that the earth has. But it has *something*. It can't avoid having an atmosphere of sorts. Here's how you can demonstrate that:

The earth is made up of two sections of radically different compositions (like an egg which is made up of a central yolk and a surrounding white). The "yolk" of the earth is the nickel-iron core, with a high density roughly about 10 times that of water. Around it, the "white" of the earth is the silicate crust, with a low density about 3 times that of water. The overall density of earth is between those two figures. It comes to 5.5 times that of water (or 5.5 grams per cubic centimeter).

The moon's density is 3.3 grams per cubic centimeter. In order to be that much less dense than the earth, the moon must be lacking a sizable nickel-iron. It must be all "white," so to speak, and consist chiefly of silicates.

It is reasonable to suppose that the elementary composition of the moon is about the same, then, as that of the earth's rocks. The two were formed at the same time out of the same materials. The earth's crust, for instance, is about 2½ percent potassium and we can assume the same figure for the moon as a whole.

The mass of the moon is about 73,430,000,000,000,000,-

This was first published in *Venture Science Fiction*, March 1958. Copyright © 1957 by Mercury Press, Inc.

000,000 kilograms, or roughly 80 quintillion tons. The mass of potassium in the moon can be placed (as a first approximation) at 1,800,000,000,000,000,000,000, kilograms or 2 quintillion tons.

There are three varieties of potassium atoms in existence. Two of them, potassium-39 and potassium-41, make up just about 99.99 percent of the total. However, 0.0119 percent of the total consists of the rare isotope, potassium-40, and that is the interesting one. The total mass of potassium-40 on the moon would come to 214,000,000,000,000,000 kilograms or 235 trillion tons.

The unusual thing about potassium-40 is that it is radioactive. It has a half-life of 1.3 billion years, which means that in that period of time, half the atoms of potassium-40 in existence break down. Most of the atoms that break down (89 percent to be exact) give off an electron and become stable atoms of calcium-40. The nuclei of the remaining 11 percent, however, take up electrons from the surrounding environment and become stable atoms of argon-40.

Once the half-life of a radioactive substance is known, its rate of breakdown per unit time can be easily calculated. On the moon, 3,600 grams (or just about 8 pounds) of potassium-40 are breaking down each second. As a result of that breakdown, 3,240 grams (7 1/6 pounds) of calcium-40 and 360 grams (5/6 of a pound) of argon-40 are formed each second.

It is the argon-40 you want to keep your eye on, since argon is a gas that means the moon is constantly forming an atmosphere of its own. Of course, 360 grams of argon aren't much, but if that much is formed each second and if the seconds keep piling up . . .

Furthermore, there was more potassium-40 present in the moon (and everywhere) in the past than there is now. About 1.3 billion years ago, there was twice as much and four billion years ago there was eight times as much as there is today.

If we calculate the amount of argon formed over the four billion years that the moon has been a solid body and make allowance for the greater quantity of potassium-40 present in the past, it turns out that the total quantity of argon formed in all that time is 150,000,000,000,000,000 kilograms, or about 170 trillion *tons* of argon.

To give you an idea how large a figure that is, it represents nearly three times as much argon as is present in our own atmosphere (which argon, incidentally, was and is being formed from our own potassium-40).

If all that argon were present on the surface of the moon, our satellite would have an atmosphere with a mass 1/30 that of ours. Furthermore, since the moon's surface is only 1/16 that of the earth, its atmosphere would be crowded together until it were just about half as dense as that of earth.

But the moon has no such atmosphere as we very well know. What, then, has happened to the moon's argon?

First the potassium-40 is spread throughout the volume of the moon. Argon that happens to be formed in the outer layers of moon-rock can make its way to the surface, but argon formed deep down would be trapped. (This is true of the argon formed on earth, too. The quantity of argon trapped within the earth is certainly five times as great and may be as much as fifteen times as great as the amount in the atmosphere.)

But even if only 1/15 of the moon's argon reached the surface, the moon would still have an atmosphere with a density some 3 percent that of the earth's atmosphere and even that much is out of the question.

Another point arises. The moon's gravitational field is only 1/6 that of the earth and it is simply not powerful enough to hold on to the argon. The moon loses its argon to outer space almost as quickly as that argon comes leaking out of the moon's rocks.

Almost! It takes a certain time for the argon to make its way clear of the moon, so that there is always *some* argon (not much) present in the neighborhood of the moon's surface.

As a matter of fact, astronomers observing the radio waves emitted by various celestial objects have studied the behavior of those waves that happened to skim the surface of the moon on their way to earth. Those radio waves were slightly disturbed and it was calculated that the disturbance was due to a lunar atmosphere of charged particles equal in density to a ten-trillionth of our atmosphere.

Not much—but something.

Chapter 15   Man and the Sun

The Sun was like a god to ancient man. Ikhnaton, pharaoh of Egypt from 1375 to 1358 B.C., worshipped the Sun and wrote a hymn to it that survives today. Fifteen centuries later, when Christianity was rising to power in the Roman Empire, its strongest competitor was Mithraism, a cult of the Sun.

And surely, if any inanimate object is worth worshiping, that object is the Sun. It brought the progression of day and night which must have given early Man his first notion of time. It brought warmth and life to the world, and each dawn was joy as light ended, once again, the terrors of the dark. If the Sun's light grew merely pale and dim, as in the winter months, ice and death closed in. No wonder, then, that whenever its brightness was temporarily eclipsed altogether from the daytime sky, sheer panic gripped those who watched.

Modern science has intensified our realization of the manner in which we depend upon the Sun. Except for volcanic heat and nuclear reactions, all energy-sources used by man come ultimately from the Sun. The oceans are kept liquid by the Sun's heat, and the vapors drawn up by that heat are restored as life-giving rain, while the warming of the atmosphere gives us our wind and weather.

The Sun's rays supply the energy required by green plants so that they might manufacture starch out of carbon dioxide, and free oxygen out of water. In a way, the food we eat and air we breathe are the direct gifts of the Sun.

What is the Sun to which we owe so much? A ball of light is what we see, and a ball of pure and perfect light, weightless and divine, is what the ancients judged it to be. One Greek astronomer used geometric propositions to show that the Sun must be larger than the Earth itself, and that the Earth must therefore move around the Sun, but few listened to such apparent nonsense.

Eighteen centuries later, however, came the Polish astronomer, Nicholas Copernicus, who, in 1543, published a detailed analysis of the manner in which the Earth must go round the Sun, if the movements of the heavenly bodies were to be explained conveniently. After a century of debate his views were accepted. In 1610, the Italian scientist, Galileo, helped drive the decision home by detecting black spots on the Sun, a stain on its supposed perfection that helped demonstrate it to be a material body and not of some semidivine and unearthly substance.

Then, in 1683, the English scientist, Isaac Newton, worked out the theory of universal gravitation, and mankind learned of another debt to the Sun. Its gigantic body spread an enormous field of gravitational force for billions of miles in every direction. Caught in that field, the Earth circled the Sun steadily, never approaching too closely, never receding too distantly; held as securely and as gently as a child in its mother's arms.

In the light of modern science, the Sun is a globe of matter, 864,000 miles in diameter, circling on its axis once in twenty-five days. The Earth compared to it is like a very small pea to a basketball. If the Sun were a hollow shell, a million and a quarter planets the size of the Earth could be dropped into it without quite filling it. Matter is somewhat more tightly packed in the Earth, than it is in the Sun as a whole, however. It would take the material of only 333,000 Earths to make up the material of the Sun.

The smallest parts of the Sun we can see are gross and monstrous. The material in its surface layers, at a temperature of 10,000° F., turn and bubble, with portions rising and sinking to give the whole a "rice-grain" appearance. But each grain is a thousand miles across.

Swirls of matter form on the Sun's surface, like gigantic tornadoes, with strong magnetic properties. The energy spent in building up this magnetism and in producing other vast disturbances is subtracted from its own heat. The tornadoes cool down, therefore, to a mere 7000° F. This is hot by ordinary standards, but is so much cooler than the surrounding surface of the Sun, that they appear black in comparison. These are the Sunspots, first detected by Galileo.

These Sunspots, these whirlwinds in the Sun, are thousands of miles across. One of them, measured in 1947, was 52,000 miles in diameter. Three dozen planets like Earth would not have sufficed to stop up that gigantic funnel.

Sunspots appear in cycles, increasing in number from year to year till a peak is reached, during which time the Sun is broadly pockmarked. The incidence then declines, until there are years when the Sun is scarcely marred. The peaks come at eleven-year-intervals, and at those times the Sun seems disturbed in many ways.

At the peaks of Sunspot activity, for instance, the Sun is particularly active in erupting material thousands and even hundreds of thousands of miles upward against its own giant gravity. These "prominences" form gouts of bright red flame that spout or arch upward, invisible to ordinary viewing but easily apparent against the edge of the Solar globe when the glare of the Sun's disk is blocked out in modern instruments.

A natural blocking out of the light of the Sun's disk takes place when the Moon passes directly in front of it. By an odd coincidence, the tiny Moon is at just the proper distance from us to match the apparent size of the giant Sun. As the Moon passes before the Sun, therefore, the fit is almost perfect.

When this happens (unfortunately for astronomers, all too infrequently) the white-hot glare of the Sun itself is shaded, and the gently glowing outer atmosphere of the Sun becomes visible as a pearly-white collection of hazy luminous streamers. This "corona" extends outward from the Sun's disk as a very hot, but very tenuous gas. It is only in the last twenty years that rocket observations have taught us how hot it really is, for the temperature of the corona turns out to be nearly 2,000,000° F. This is hot enough to radiate X rays as well as ordinary light. The matter of the corona, however, is very thinly spread through space, and high though its temperature is, its total heat content is small.

Astronomers suspect that during the infancy of the Solar system, the matter making it up consisted of a thinly spread out mass of dust and gas that slowly swirled and contracted under its own gravitational pull.

As the matter compacted toward the center, the temperature at that center rose. This is an inevitable phenomenon. Compressing air by means of a handpump heats that air; and the center of the Earth, compressed by the weight of all the rock and metal above it, is at a temperature of thousands of degrees.

The contracting matter of the Sun, much more massive by far than that of the Earth, raised its own internal pressures and temperatures to incredible heights.

Atoms move more energetically, the higher the temperature, and a point is reached where the collisions are so fierce that the electrons that fill the outer reaches of the atoms are stripped off and the tiny nuclei at the center of the atoms are laid bare. At that point, matter collapses together drastically, and the Sun shrank rapidly to become the globe that now exists.

Almost all the matter of the early Sun was hydrogen and the nucleus of the hydrogen atom is a single, incredibly tiny particle, called a "proton." As the temperature continued to rise, these protons, now laid bare, collided with more and more force, until they began to interact to form a more complicated nucleus built up of four particles, a nucleus of helium.

This fusion of hydrogen to form helium liberates floods of energy. It is the process that goes on in a hydrogen bomb. In short, the Sun ignited to form a nuclear fire and become a huge hydrogen bomb, in the light and warmth of which we now live. The Sun does not, like a tiny hydrogen bomb on Earth, blow apart and vanish into nothingness after the few instances of explosion, because the Sun's giant gravity holds it substance together against all the force of nuclear fusion.

Nor are we subjected to the dangerous radiation of this great hydrogen bomb in the sky, because most of the danger is buried deep within the Sun. At its center, where the nuclear fusion is proceeding, the temperature is about 25,-000,000° F., but the incredible heat is contained there and can leak out but slowly through the hundreds of thousands of miles of Solar matter. The Sun's surface is only mildly warm in comparison and what dangerous radiation survives is largely absorbed by the Earth's atmosphere before it can reach us.

It has probably been five or six billion years since the compacting center of the Sun ignited into nuclear fire but in all that time only a small portion of its immense hydrogen content has been fused to helium. Even today, considerably more than half the mass of the Sun is hydrogen, and there is enough nuclear fuel left in it to keep it burning much the way it is now for at least ten billion years more.

More reaches us from the Sun than mankind suspected until recent times. The material in the prominences that blast upward from the Sun's surface does not all return to the Sun. Some (like sea-spray driven far inland by the wind) leaves the Sun altogether and spreads outward in ever-thinning wisps.

This material, in the form of the tiny electrically-charged protons and electrons, feeds the corona, which spreads out from the Sun, wider and wider, until it is lost in the vast emptiness of space—only to be constantly renewed by new matter from the Sun. This thin matter, blasting ever outward, is the "Solar wind" and it makes itself felt even at the distance of the Earth, 93,000,000 miles from the Sun.

The Solar material is exceedingly thin here in the neighborhood of the Earth, but it is still thick enough to keep the space around us from being a complete vacuum. The Earth, in other words, can be viewed as actually moving in an orbit within the Sun's corona.

The charged particles from the Sun are attracted by the Earth's magnetic field, which bellies outward from the magnetic poles in the polar regions and reaches its greatest height over the equatorial regions.

Electrons and protons from the Sun collect in the Earth's magnetic field, following its shape and forming a doughnut-like encirclement of the Earth. These are the Van Allen belts, named for the American physicist, James A. Van Allen, who discovered them in 1958.

Near the magnetic poles, the charged particles dip toward the Earth's upper atmosphere where they undergo interactions that create the eerie shifting beauty of the Northern (and Southern) Lights.

The Solar wind is not constant. Every once in a while, unpredictably, it grows more intense. This happens most often at the times of Sunspot peaks and is particularly associated with "Flares." Occasionally, the immediate neighborhood of a Sunspot may grow intensely brilliant for an hour or so, and from such a flare an enormous spray of particles is discharged into space.

If that spray happens to be in the direction of the Earth, the cloud of particles will invade our upper atmosphere in less than a day. The Northern Lights will grow bright indeed and we have what is called an "electric storm."

Such a storm can seriously affect modern technology. Radio communication depends on the fact that a region in the upper atmosphere contains electrically-charged atom fragments called "ions" (so that the region is the "ionosphere"). These ions can reflect radio waves. However, when charged particles invade the ionosphere in floods, this reflecting action becomes erratic. Mankind's long-distance electronic means of communication break down into a flood of static that may persist for thirty hours.

The Solar wind may have more everyday effects on Earth too; effects that are more intrinsically important. Rainfall is not entirely a matter of moisture in the air, or even of clouds, as we now know. Raindrops must somehow be made to form and this is not always easy. Raindrops usually form about some dust particle of the correct size, shape and chemical properties. The modern rainmaker tries to supply such dust by spraying appropriate chemicals into clouds.

Ions also form natural nuclei for raindrops and it may be that the probability of rainfall, all things being equal, depends on just how rich in ions the upper atmosphere is. On the whole, the ions are more numerous in the years when Sunspot activity is high and the Solar wind is strong. Rainfall may therefore tend to be higher in those years.

Thus, some measurements have indicated that the water level in Lake Erie is higher at Sunspot maximum. Studies of the rings of trees in southwestern United States seem to show that the rings are thickest (and rain, therefore, most copious) in an eleven-year cycle, like that of the Sunspots.

When we stop to think of how thoroughly all of life can be affected by variations in rainfall, we might blame almost anything on Sunspots. Periodic shortages of rain might mean periodic decreases in food supplies and therefore periodic times of political unrest and therefore periodic aggression abroad. It is no wonder that some people have tried to work out cycles of wars and depressions to match the rise and fall of Sunspot frequency. However, the variation in Sunspot frequency is sufficiently irregular, and human behavior is sufficiently complex, to make such attempts futile, at least so far.

In the coming age of space travel, the Sun's behavior will be of immediate concern to astronauts. The Earth's atmosphere absorbs much dangerous radiation and outside that atmosphere the safety margin grows thin. As long as astronauts remain in the immediate neighborhood of the Earth for short periods of time, the walls of the space capsule (and, even more important, Earth's magnetic field) will protect him, but farther out, the danger sharpens.

On the way to the Moon, he must be protected against the intense radiation of the Van Allen belts. Perhaps he will have to avoid them altogether by being aimed through the polar gaps in the belts.

Out in open space, the astronaut cannot count on safety even under conditions when the ordinary radiation level in his surroundings would seem to be satisfactorily low. A sudden flare on the Sun's surface may spray dangerous particles in his direction, which may be impossible to ward off. Some flares have proven so fierce as to send out quantities of the most energetic radiation known—cosmic rays.

Explorers on the Moon itself, which has no atmosphere to speak of (see Chapter 14), may find that one of the major dangers against which they must guard will be the erratic behavior of the Solar wind, with its occasional and unpredictable spurts of death.

Clearly, it would be wise to know more about the Sun in every way. One exciting development which may teach us much involves a strange little particle called the "neutrino." The fusion reactions going on in the Sun's center liberate these particles as well as ordinary radiation.

Ordinary radiation takes so long to get to the Sun's surface and undergoes so many changes in the process, that what we finally see of it tells us only about the Sun's surface and nothing at all about its interior, except what we can deduce indirectly.

The neutrinos, however, are so tiny, and so indifferent to ordinary matter, that they tear out of the center of the Sun at the speed of light without being affected by the substance of the Sun in the slightest. They reach us eight minutes after being formed, coming straight from the Solar center.

Scientists are now designing "neutrino telescopes" which can consist, for instance, of large tanks of certain chemicals that may be able to stop just a few of the neutrinos pouring out of the Sun. From the number that are stopped, and from other information that can be gained from them, it may be possible to work out the temperature and other conditions at the Sun's center, with a certainty far superior to anything we can now boast.

With the very center of the Sun exposed, much that is mysterious now may become mysterious no longer. Sunspots, prominences, flares, the Solar wind, may all be charted in detail and, perhaps, in advance. With this new knowledge, we may be able to guide ourselves safely across the vast stretches of space, as once the compass guided the European explorers over the terrible wastes of open sea.

Chapter 16   The Unused Stars

There is the familiar story of the wide-eyed young thing who attended a lecture on popular astronomy and who said afterward, "I understand about how astronomers find out the distance of the stars and how hot they are. What gets me, though, is how they found out their *names*."

Actually, very few stars have real names. Most are known by their listing in some catalogue and in place of names own a string of digits.

Even many of those bright enough to be seen by the unaided eye are known chiefly by Greek letters applied to the name of the constellation in which they are to be found.

This article appeared in *Amazing Stories*, July 1959. Copyright © 1959 by Ziff-Davis Publishing Co.

The nearest star, Alpha Centauri, is so-called because it is the brightest star in the constellation Centaurus and therefore deserves to be tabbed by "Alpha," the first letter of the Greek alphabet. Alpha Centauri means "the first of the Centaur." There are also Beta Centauri, Gamma Centauri, and so on.

However, there are about 250 stars with names all their own: real names, mouth-filling names, of which perhaps not more than a dozen are known to the general public. This is a pity for surely there would be something pleasant to say about a star named Ruchbah, or another named Benetnasch. Those are real names of real stars.

Even the old reliables, the few star names that are used so often that even nonastronomers have heard of them, gain new vitality if we consider what the names mean.

The brightest star in the heavens and the one bearing the most familiar of the real names is Sirius. This is in the constellation Canis Major ("Big Dog"—the official names of all the constellations are in Latin) and is sometimes called the "Dog Star" for that reason. Because it is so bright a star, the ancients had the sneaking suspicion that it added significantly to the sun's heat, when it rose with the sun in midsummer. We still call midsummer the "dog days" and the name Sirius itself may have arisen from that line of thought since it comes from a Greek word meaning "scorching."

(Incidentally, Sirius, the Dog Star, has a companion star that is extremely small, just a little more than twice the diameter of our tiny earth. This companion is sometimes referred to irreverently as "The Pup.")

A bright star to the west of Sirius is in Canis Minor ("Little Dog"). Since it is to the west of Sirius, it naturally rises and sets a little earlier than Sirius. This star, rising before the Dog Star does, is called Procyon, from Greek words meaning "before the dog."

Near the two constellations Ursa Major ("Big Bear") and Ursa Minor ("Little Bear") is the constellation Boötes ("Herdsman"). The ancients pictured the constellation as a man holding two dogs in leash. The dogs were represented by stars in a small constellation between Boötes and Ursa Major, these being Canes Venatici ("Hunting Dogs"). Boötes

152

and the dogs were obviously protecting the rest of the heavens against the ferocious bears. Consequently, the brightest star in Boötes was named Arcturus, from Greek words meaning "guardian of the bears."

The ancients took the imaginative pictures they drew seriously. For instance, the constellation Auriga ("Charioteer") was drawn by them as an old man holding a bridle in one hand and a goat and her kids in the other. The stars at the side of the constellation are therefore referred to as "The Kids" and the brightest of these (and of the whole constellation) is Capella, from a Latin word meaning "Little Goat." Capella is often called the Goat Star for this reason.

Again, the constellation Virgo ("Maiden") is pictured as a young woman holding ears of grain in her hand. Presumably, this is because the sun enters Virgo in the early fall when the grain is ripe and ready for harvest. The star in those ears of grain is Spica, the Latin word for "ear of grain."

Sometimes the names are less dependent on the constellation pictures. The constellation of Gemini ("Twins") contains two bright stars closely spaced in the heavens (which probably inspired the name of the constellation). The Romans called them after the most famous twins in their mythology, Castor and Pollux.

Regulus is the brightest star in Leo ("Lion") and comes from the Latin word meaning "Little King" which is appropriate for the chief ornament of the king of beasts. More appropriate still is Antares, which means "Rivaling Ares." Ares is the Greek god whom the Romans called Mars. Antares is a red star rivaling Mars in color.

Most appropriately named of all is Polaris, the star marking the North Celestial Pole, which is also called the North Star and the Pole Star.

Then there are completely inappropriate names. The constellation Orion ("Hunter") is pictured as a giant who is holding up his left hand to ward off the onrushing Taurus ("Bull") while he is ready to strike with the club in his right hand. Bellatrix is the star in his left shoulder and its name is the Latin word for "female warrior," something I imagine Orion would resent.

However, the large majority of the star names are neither Greek nor Latin but are Arabic (hence the number of stars with names beginning with "Al-," the Arabic word for "the").

Consider, for instance, the seven stars of the Big Dipper. Everyone has seen them; they are the one group of stars that anyone in the northern hemisphere can point out, even if they know nothing else about the stars. But what are their names?

To be sure, many of us have learned to call the two stars at one end of the group that form a line aimed at the North Star, "the Pointers" but what are their real names?

Well, here are the names, starting at the end of the handle of the Dipper and ending with the Pointers: Alkaid, Mizar, Alioth, Megrez, Phecda, Merak, and Dubhe.

The first star in the list, Alkaid, has a name that sounds as though it were meant for an antacid tablet, but it is Arabic (as all the names are) and means "the leader" since it leads the seven stars across the sky.

The second star, Mizar, means "veil." Behind that, lies a story. Near Mizar is a considerably fainter star. If this second star were by itself it could be seen without trouble, but the presence of the nearby brighter star veils it, you see. In order to distinguish the weaker star, you must have good eyesight, and for centuries this star-pair was an eye chart in the sky and was used to distinguish good sight from bad. The fainter star is Alcor, from the Arabic for "weak one."

The name of the third star in the Big Dipper, Alioth, is an Arabic word for the fat tail of a sheep. If that sounds startling, you'll have to realize that the Greeks pictured the Big Dipper as the Great Bear in such a way that the four stars forming the bowl of the dipper formed his rear half, while the three stars of the handle of the dipper formed a tail. Now you and I know that bears have no tails to speak of, and the Greeks must have known that, too. They must have inherited the pictures from the Babylonians and named the animals bears despite the tails. The Arabs too were stuck with the tails, and since they had no word for the tail of a bear, they called the star by the name of the tail of a sheep.

The fourth star, which begins the bowl of the dipper, is Megrez, meaning "root," presumably because it is the root of the tail.

I have not found out the meaning of Phecda, but as for the pointers, Merak (the further from Polaris) means "loin" since it is located in the loin of the bear, while Dubhe just means "bear."

Similarly, the four stars of the famous "Square of Pegasus" (Pegasus is the famous "Flying Horse") have the Arabic names of Alpheratz, Algenib, Markab, and Scheat. Alpheratz, in the flank of the horse, means "the mare"; Algenib, higher up, is "the side"; Markab, still higher up, is "the saddle." Scheat, just above a foreleg, is not as clearly named. It may be derived from the word for "good fortune," but if so, there is no obvious reason for it.

A number of the more familiar star names are also of Arabic origin. The second brightest star in Orion, the one in the left leg of the pictured hunter, is Rigel, from the Arabic word for "foot." Betelgeuse, the brightest star in the constellation, is in the uplifted right arm of the hunter and is a corruption of an Arabic phrase that originally stood for "arm of Orion."

Other stars are also blessed with names literally derived from the pictures of the constellations to which they belong. Altair, the brightest star in the constellation Aquila ("Eagle") means in Arabic "the bird." The constellation Pisces ("Fishes") is pictured as two fish held together by a long cord. In the middle of the cord is the brightest star of the constellation to which the Arabic astronomers gave the name Al Rischa ("the cord").

The brightest star in the constellation Cygnus ("Swan") is Deneb. It is located in the rear portion of the swan as usually pictured and comes from an Arabic word meaning "tail." This was a favorite star name among the Arabs (who were the great astronomers of the early Middle Ages, which is why so many stellar names hark back to them) so that there are a number of Denebs in the sky. The Arabs distinguished among them by adding a second word for the constellation. This persists in some cases. For instance, Deneb Algedi in Capricornus ("Goat") means "tail of the goat," and Deneb Kaitos in Cetus ("Whale") means "tail of the whale." The second brightest star in Leo is Denebola, the "-ola" suffix being what is left of that portion of the Arab phrase meaning "of the lion."

On the other hand, just to show that the Arabs are not restricted to one end of the creature, the brightest star in Piscis Australis ("Southern Fish") is Fomalhaut, from an Arabic phrase meaning "mouth of the fish." Similarly, the brightest star of Ophiuchus ("Serpent-Holder"), pictured, naturally, as a man holding a serpent, is Rasalhague, meaning "head of the snake-charmer."

Aldebaran, the brightest star in Taurus, is a kind of Procyon in reverse. Aldebaran is a little to the east of the well-known group of stars called the Pleiades and consequently follows them both in rising and in setting. The name of the star means "the follower" in Arabic.

Perhaps the most colorful Arabic name for a star is that for the second brightest star in the constellation Perseus. It is one of the few bright stars in the sky that changes brightness visibly and regularly. This was a startling thing to the ancients, who generally believed that the stars were perfect and unchangeable. The case of this star may have therefore guided the picture they formed of the constellation. This shows Perseus holding the severed head of Medusa, a horrible demon whom he had killed—a demon so horrible, with an awful face and living snakes in place of hair, that she froze men into stone with a mere look.

The star in question is right in the head of Medusa and the Arabs named it Algol, meaning "the ghoul." Algol is consequently known as the "Demon Star."

All this just gives an idea of the richness of the heavens. Among the two hundred or so names I have not mentioned are such succulent samples as Tarazed, Pherkad, Mesartim, Kochab, Izar, Caph, Dschubba, and Azelfafage.

Chapter 17   Measuring Rods in Space

Mankind, for convenience's sake, makes use of units of different size to suit different needs. The length of a room is measured in feet, of a racetrack in yards or furlongs, of an automobile trip in miles.

This is done chiefly in order to keep the number of digits involved at a reasonable quantity. It would be ridiculous to say that a room was 0.0038 miles long, rather than 20 feet; or that the distance from Boston to New York was 1,200,000 feet rather than 230 miles.

However, none of the common units invented for use on earth's surface is particularly convenient for the measurement of astronomical distances. The longest unit of length on earth that is in common use is the mile, or, in countries that use the metric system, the kilometer, and both are far too short for astronomers. (A mile is equal to 1.61 kilometers.)

The nearest object to us in space is the moon and the second nearest, of any considerable size, is Venus. But the average distance of the moon from earth, expressed in our own customary units, is 237,000 miles (380,000 kilometers) while Venus comes no closer than 25,000,000 miles (40,-000,000 kilometers).

To keep numbers from rising into such uncomfortable millions, billions, and beyond, astronomers have for a long time been using larger units of measurement not so familiar to earth-bound mortals. But now that the Age of Space is here, these units are bombarding all of us with increasing frequency. We must learn to understand this astronomical code of distance.

For instance, astronomers use the distance between the earth and the sun as one space yardstick. This varies by several million miles according to the exact position of the earth in its elliptical orbit, but the average distance is 92,870,000 miles (149,450,000 kilometers).

Astronomers call this measuring rod the Astronomical

This appeared in *Space World*, September 1961.

Unit, often abbreviated A.U. You can say that the average distance of the earth from the sun is 1 A.U. The advantage of doing this is that you can measure other astronomical distances in A.U. and thus work with convenient terms that are quickly understandable.

For instance, the moon's average distance is 0.00255 A.U., while Venus is within 0.27 A.U. of the earth. This tells you at once that the lunar distance is 1/400 that of the sun and that Venus is at ¼ the solar distance.

Table 1 gives the mean distances of the various planets from the sun in miles, kilometers, and A.U. Not only are the numbers in the A.U. column easier to handle, write, and say, but they also make clear instantly the relationship between the various values, as the mile and the kilometer measurements cannot readily do.

If you are told that Neptune is 2,800,000,000 miles away from the sun, you have nothing more than a rather confusing number. If, on the other hand, you are told this distance is 30.07 A.U., you know at once that Neptune is 30 times farther from the sun than earth is.

TABLE 1

Planet	Distance in Miles	Distance in Kilometers	Distance in A.U.
Mercury	36,000,000	58,000,000	0.39
Venus	67,500,000	108,300,000	0.72
Earth	92,870,000	149,450,000	1.00
Mars	142,000,000	228,000,000	1.52
Jupiter	484,000,000	778,700,000	5.20
Saturn	888,000,000	1,428,000,000	9.54
Uranus	1,790,000,000	2,872,000,000	19.19
Neptune	2,800,000,000	4,500,000,000	30.07
Pluto	3,700,000,000	5,900,000,000	39.46

From the A.U. figures, you can tell at a glance that Saturn is almost twice as far from the sun as Jupiter and that Pluto is (on the average) twice as far away as Uranus. The same information is available from the mile and kilometer columns, but these large, unwieldy numbers make the facts far less obvious.

A measuring rod that is of major importance to astronomers involves the velocity of light.

In one second, light (or any other form of electromagnetic radiation such as radio waves) travels 186,282 miles. Rather than call this "the distance light travels in one second" we can leave out most of the words and simply use the term "light-second." How easy to say, then, that the moon's mean distance from earth is 1.27 light-seconds, or that Venus approaches us within 13.5 light-seconds.

It is practical to do so, too, for when we establish radio contact with an exploring expedition on the moon someday, it will take 1.27 seconds for our signal to reach them. A radar signal bounced off Venus at its closest distance of 13.5 light-seconds will take 27 seconds for the round trip. Distance measured in such units will obviously fit neatly with the timing of radio communications.

The solar system can be measured off in light-seconds but this would be more unwieldy than using Astronomical Units. One A.U. is equal to just about 500 light-seconds. Consequently, the mean distance of Neptune from the sun, which is about 30 A.U., would be 15,000 light-seconds. The latter is a larger and clumsier figure, and therefore less convenient.

However, the use of light's velocity as a measuring rod is not limited to light-seconds. There is the distance that light can cover in a minute or in an hour (the "light-minute" and "light-hour"). Naturally, a light-minute is equal to 60 light-seconds, while a light-hour is equal to 60 light-minutes or 3600 light-seconds.

In Table 2, the mean distances of the planets from the sun are given again, this time in light-minutes and light-hours. As you see, the light-minute can be a convenient measuring rod for planetary distances as far out as Jupiter's orbit, while light-hours are more convenient for the planets beyond.

The width of the known solar system from one end of Pluto's orbit to the other is about 11 light-hours, or almost half a "light-day." Outside that stretch lies nothing we know of (except for the insubstantial ghosts we call comets plus some wandering meteors, perhaps) until we reach the stars.

TABLE 2

Planet	Distance in Light-Minutes	Distance in Light-Hours
Mercury	3.2	0.053
Venus	6.0	0.10
Earth	8.3	0.14
Mars	12.7	0.21
Jupiter	43.3	0.72
Saturn	78.6	1.31
Uranus	159	2.65
Neptune	250	4.18
Pluto	330	5.50

A graphic picture of our family of planets now forms in our minds. Light, which can streak from earth to moon in 1¼ seconds and can reach us from the sun in approximately eight minutes, must labor steadily onward for eleven long hours in order to span the far-stretching orbit of Pluto.

Yet the solar system is but a speck in the vastness of space and astronomers probe far beyond its limits with their telescopes. Fortunately, the velocity of light continues to offer them a succession of longer and longer measuring rods. However, if we suppose that "light-weeks" and "light-months" would be handy as the next units, we are mistaken.

This is because light waves, having passed the bounds of Pluto's orbit, can travel on for weeks and months in any direction and meet with nothing substantial that we know of in the near-perfect vacuum of outer space.

There is no known object out there whose distance from our sun can be conveniently measured in terms of light-weeks or even light-months.

By the time the nearest star is reached, we will have graduated to the "light-year"—186,282 miles multiplied by the number of seconds (over 31 million of them) in one full year. Such a distance is long indeed; 5,890,000,000,000 miles or 9,450,000,000,000 kilometers. Roughly speaking, a light-year is equivalent to nearly six trillion miles or nearly ten trillion kilometers.

Great as that distance is, there is no known body outside the solar system that is within a light-year of us. The nearest star, Alpha Centauri, is 4.3 light-years away.

A second measuring rod useful in measuring the distance of stars is not based on the velocity of light. Instead, this new unit involves the apparent shift in position of the nearer stars against the background of more distant stars. This shift comes about as the earth travels from one point in its orbit to the extreme opposite point six months later. Half this apparent change in the star's position is called the "stellar parallax."

You can observe a crude type of parallax if you hold your finger up about six inches before your nose and sight it against some distant object with only one eye open. Now, without moving your finger, sight with your other eye. The finger shifts position against the background because you have shifted the point of view from one eye to the other.

The more distant the object you are sighting, the smaller the parallax. Extend your finger to arm's length and you will notice how much less it shifts against the background as you switch from eye to eye. For this reason, it is possible to calculate the distance of a celestial object from the size of its shifting parallax. This method was used to estimate the distance of the closest stars more than a century ago, but it was a difficult task, for even nearby stars have extremely small parallax shifts.

Imagine a parallax of one second of arc (which is 1/60 of a minute of arc, which in turn is 1/60 of one degree of arc out of the 360 degrees that make up the full circumference of a circle). A second of arc is equal to the apparent diameter of a one-cent piece held at a distance of 2½ miles, so it is exceedingly tiny. A star with this parallax would be a "parsec" away. (The word parsec is a shorthand contraction for "*par*allax of one *sec*ond.")

But even such a tiny parallax is too large. No known object outside the solar system is as close as a single parsec.

Consequently, no known star has a parallax as great as one second of arc. The nearet star, Alpha Centauri, has a parallax of 0.76 second of arc.

As it turns out, a parsec is equal to 3.26 light-years. Thus, Alpha Centauri, 4.3 light-years away, is 4.3 divided by 3.26 or 1.3 parsecs distant.

In Table 3, the distance of some of the more familiar stars is given in both light-years and parsecs.

You might now think that astronomers have all the measuring rods they need, but actually, the stars listed in Table 3 all belong to our immediate stellar neighborhood, a small section of a particular spiral arm of our galaxy. The whole Milky Way is far larger than the small volume which holds all the suns we can see with the naked eye—including those stars whose parallaxes are listed.

TABLE 3

Stars	Distance in Light-Years	Distance in Parsecs
Alpha Centauri	4.3	1.3
Sirius	8.6	2.6
Procyon	11	3.4
Altair	16	4.9
Fomalhaut	23	7.1
Vega	27	8.3
Pollux	33	10
Arcturus	40	12
Capella	42	13
Castor	45	14
Aldebaran	55	17
Regulus	77	24
Canopus	100	31
Mira	165	51
Antares	220	68
Betelgeuse	275	85
Deneb	400	120
Rigel	540	165

The nucleus, or core, of our galaxy, which contains about 90 percent of all its stars (and which we can't even see by an optical telescope because of obscuring dust clouds between it and ourselves), is no less than 30,000 light-years from us. In fact, the diameter of our disc-shaped galaxy is about 100,000 light-years altogether, while its greatest thickness (at the center) is about 30,000 light years. As you see, the numbers are creeping upward again.

One way to prevent this numerical growth is to use a "light-century" (100 light-years) and a "light-millennium" (1,000 light-years or 10 light-centuries). Then we could say that a star like Deneb was 4 light-centuries distant from us and that the measurements of the galaxy were 100 light-millennia across by 30 light-millennia thick.

Actually, these units are rarely used. Instead, astronomers tend to favor the parsec as the base of long measurements. Just as, in the metric system, a kilometer is equal to 1,000 meters and a kilogram to 1,000 grams, so astronomers have set up the "kiloparsec" equal to 1,000 parsecs. Using this compact measuring rod, we would say that the measurements of the galaxy were, roughly, 31 kiloparsecs across by 9 kiloparsecs thick.

However, even our galaxy is but a dot in the enormity of cosmic space filled with other galaxies—myriads of them. Nearest to us are the comparatively small "satellite galaxies" of our Milky Way—the Large Magellanic Cloud and the Small Magellanic Cloud. These are, separately, 150,000 and 170,000 light-years away, or 47 and 53 kiloparsecs.

The nearest large galaxy to ours is the Andromeda galaxy which is 2,300,000 light-years away or 700 kiloparsecs. Other galaxies (including, for instance, a famous galactic cluster in the constellation of Coma Berenices, and a spectacular galaxy in Cygnus which has been thought to be two galaxies in collision) are much more distant. In considering these distant galaxies, even the kiloparsec is too small a measuring rod.

Instead, suppose we use the "megaparsec," equal to a million parsecs or to a thousand kiloparsecs (or to 3,260,000 light-years). Using this unit, the cluster of galaxies in Coma Berenices is 25 megaparsecs distant. The colliding galaxies in Cygnus are 80 megaparsecs distant.

Do we have a measuring rod now that will need no further enlargement? No, not quite. In 1963, astronomers came to realize that there were objects in the universe that seemed to be far more distant than even the most distant ordinary galaxies. These new objects, the most distant known, are called quasars (see Chapter 19).

The most distant quasar yet detected is one called 3C9 which is thought to be possibly 9 billion light-years distant. Such a distance is 2800 megaparsecs.

Suppose, then, we go one step farther and introduce the "gigaparsec," which is equal to a billion parsecs or a thousand megaparsecs. We could say then that 3C9 is 2.8 gigaparsecs from us.

In fact, astronomers have reason to think that the farthest distance we can reach out with any instrument, however advanced and perfect, is 12.5 billion light-years. If so, then the width of the entire universe reachable by instruments under the best circumstances is 25 billion light-years or just about 7½ gigaparsecs.

And we need go no farther than that.

Chapter 18   Time-Travel: One-Way

In 1905, Albert Einstein advanced a new way of looking at the universe that seemed to transcend and subvert "common sense." It seemed a weird outlook indeed, one in which objects changed as they moved, growing shorter and more massive. In the new outlook, what one person would see and measure and swear to, another person would not. All our most cherished certainties seemed to dissolve.

The only consolation the average man could count on was that under ordinary circumstances, the new changeability was so small it could be ignored.

Suppose we begin, for instance, by constructing an imaginary freight train which, when standing still, is exactly one mile long and exactly one million tons in mass. If it were to chug past us at sixty miles an hour, and if we could

This article appeared in *The North American Review,* Summer, 1964.

make the necessary measurements accurately enough while it was moving, we would find that it had shortened by a ten-billionth of an inch and had become more massive by a hundred-thousandth of an ounce.

A person on the freight train, however, making the same measurements, would find the length and mass of the train unchanged. To him it would still be exactly one mile long and exactly one million tons in mass. In fact, as far as the man on the train was concerned, we ourselves (the observers watching the train) would be the ones who were slightly distorted in shape and mass.

But who in blazes is going to argue over billionths of inches and ounces? It might seem that a complicated new view of the universe involving such insignificant changes is scarcely worth the trouble.

Yet the changes are not always insignificant. Just a few years before Einstein had advanced his theory, it had been found that radioactive atoms were shooting out tiny sub-atomic particles that traveled at velocities far greater than that of our imaginary freight train. The velocities of the sub-atomic particles were anywhere from 10,000 to 186,000 miles *per second*. For them, length and mass changed drastically; changed enough to notice and measure; changed enough so that it was impossible to ignore the matter. The old notion, then, of a universe in which length and mass were unaffected by motion had to be abandoned. Einstein's outlook had to be adopted instead.

Of course, if we imagine freight trains, or anything else, taking up velocities so great as to make changes in length and mass really noticeable, they will escape from the earth's gravitational field at once. We would find ourselves in outer space and, since that is so, we might as well imagine ourselves out there to begin with.

Let's imagine ourselves on a spaceship named A, which is 1,000 feet long and has a mass of 1,000 tons. Passing us, at 162,000 miles per second, is the sister ship, B, which was built to have the same length and mass as A.

As B passes us, we use some sophisticated device to measure its length and its mass and we find that it is only

500 feet long and has a mass of 2,000 tons. It has, in other words, halved its length and doubled its mass.

We at once radio B and tell them this, but B informs us that by their own measurements, it has not changed at all. In fact, as they passed us, they measured us and found that it was *we*, A, that was only 500 feet long and fully 2,000 tons in mass.

The ships change course, approach, and rest side by side. Measurements are made and both ships are found to be normal. Both are 1,000 feet long and 1,000 tons in mass.

Which set of measurements was correct? The answer is that all were. Measurements, remember, change with motion. To the crew of A, it seemed that B was flashing past in the forward direction at 162,000 miles per second; and to B it seemed that A was flashing past in the backward direction at 162,000 miles per second. Each observed the other moving at this particular velocity and each measured the other as half-length and double-mass. Once the ships were side by side, however, each would consider the other motionless and the measurements would revert to "normal."

If you still insist on asking, "But *did* ship A shorten or didn't it?" then consider that in making a measurement, you are not necessarily checking "reality." You are merely reading the setting of a pointer, and this setting can vary under different conditions.

Einstein's theory involves more than length and mass; it involves time as well. According to Einstein, everything on a moving object slows down. The pendulum of a clock in motion moves more slowly; the hairspring of a watch pulsates in more leisurely fashion. All motion slows.

But it is periodic motion that we measure time by; some regular vibration, pulsation, or beat. If all these motions by which we measure time slow down, then we have every right to say that time itself slows down.

To some people, this seems harder to swallow than Einsteinian changes in length and mass. Length and mass are, after all, changeable in some ways. We can make an object shorter by hammering it; we can make it lighter by letting some of its water content (assuming it has some) evaporate. But nothing we know can change the rate at which time moves. We take it for granted that the time rate is something immutable; something which, oblivious to all things, proceeds unalterably on its way.

And yet, Einstein's postulated change in time rate with motion has actually been measured. Even with velocities of a few inches per second, a physical phenomenon known as the Mössbauer effect (after its discoverer) enables us to measure the excessively minute changes in time rate. Again, though, subatomic particles offer us velocities great enough to make the change easily measurable and quite significant.

There is a particle called the mu-meson which lasts for two microseconds (a microsecond is a millionth of a second) before breaking down. At least, it lasts two microseconds if it is moving at moderate speeds. Sometimes, however, a mu-meson is formed high in the atmosphere by cosmic rays and, in the shock of creation, comes streaking downward toward earth's surface at a velocity of over 180,000 miles a second.

If the mu-meson retained a lifetime of two microseconds at that speed, it ought to have time to move only about 1700 feet. Since it is formed miles high in the atmosphere it should, therefore, never last long enough to reach us here on the earth's surface.

But it *does* reach us. A really fast mu-meson can travel three miles or more before breaking down. This can be explained by supposing that time slows down for it. It still lives two microseconds by its own reckoning, but these are now (according to an earth-bound observer) very slow microseconds that stretch out over twenty ordinary microseconds.

The change in time rate exhibited by the mu-meson exactly fits Einstein's prediction, and so we must accept time as not an immutable thing, but as something with properties that depend on one's point of view.

Let's return to our spaceships A and B again, then. Once more, we suppose B to be flashing past A and we can further imagine that there is an instrument on board A which enables its crew to observe a clock on B for exactly one hour by A's clock.

The clock on B will seem slow to the observing crew, because B is moving. After one hour by A's clock, the clock on board B will have recorded a bit less than an hour. The faster B is moving, the slower its time rate and the less time will be recorded by B's clock as having elapsed.

There is a formula that can be used to work out the slowing of time rate with motion and its use yields the following table:

Velocity of B with Respect to A (miles/second by A's clock)	Time Elapsed on B's Clock After 1 Hour by A's Clock
1,000	59 min. 50 sec.
50,000	57 min. 47 sec.
100,000	52 min. 18 sec.
120,000	45 min. 54 sec.
140,000	39 min. 36 sec.
160,000	30 min. 40 sec.
170,000	24 min. 25 sec.
180,000	12 min. 13 sec.
185,000	7 min. 48 sec.
186,200	1 min. 50 sec.
186,282	no time at all.

And what happens if B travels past A at a velocity greater than 186,282 miles per second? Does its clock register less than no time? Does it start going backward?

No! We can avoid the possibility of time traveling backward, for 186,282 miles per second is the maximum possible relative velocity that can be measured. That is the velocity of light in a vacuum and, according to Einstein's theory, that relative velocity cannot be exceeded by material objects.

But there is one thing we must not forget. The crew on A observes B flashing past in the forward direction, but the crew on B observes A flashing past in the backward direction. To each crew, it is the other ship that is moving. So if the crew of B measured the clock on board A they would find *that* clock, A's clock, to be running slow.

This is serious, much more serious than the disagreement

on length and mass which was mentioned earlier. To be sure, if two ships got together after a length-mass experiment we could imagine their crews arguing.

"When you passed us, you were shorter and heavier than I was."

"No, no, when *you* passed *me*, *you* were shorter and heavier than *I* was."

"No, no—"

An argument such as this can't be settled and doesn't have to be. If an object shrinks to half its length and then returns to normal or if it doubles in mass and then returns to normal, its adventure leaves no mark. There is no trace left behind to show whether it shrank temporarily or not, or whether it grew temporarily heavy or not. Arguments over that are futile and therefore unnecessary.

But if the clock on one ship is running slower than the clock on the other ship, then, when the two ships get together, *the clocks ought to bear the record of that.* If the two clocks were synchronized at the start of the experiment, they should no longer be synchronized at the end.

Let's say that one clock, because of the slowing of its time rate, lost a total of one hour. Therefore, when the ships come together again, one clock should say 2:15 if the other says 3:15.

But which clock says which time? The crew on A swears that the clock on B was slow, while the crew on B swears just as vehemently that the clock on A was slow. Each group of men fully expects that the other clock will be one hour behind their own clock. Since both can't be right, this seems an insoluble dilemma, one that is commonly called the "clock paradox."

Actually, it isn't a paradox at all. If one ship just flashed by the other and both crews swore the other ship's clock was slow, it wouldn't matter which clock was "really" slow, because the two ships would separate forever. The two clocks would never be brought to the same place at the same time in order to be matched, and the clock paradox would never arise.

On the other hand, suppose the two ships *did* come together after the flash-past so that the clocks *could* be compared. In order for that to happen, something new must be added. At least one ship must accelerate; that is, change its velocity. If it is B that does so, it must travel in a huge curve, point itself back toward A and then slow down to the point where it could hang motionless next to A.

The action of acceleration spoils the symmetry of the situation. B changes its velocity not only with respect to A but also with respect to all the universe, all the stars and galaxies. The crew on B might insist their ship is remaining motionless and that it is A that is somehow moving and approaching them, but then they must also say that the entire universe is changing position with respect to their ship. The crew on A, however, sees only B change its velocity; the universe remains unchanged in velocity relative to A.

It is because B accelerates with respect to the entire universe (not merely with respect to A) that brings about a slowing of B's clock of a kind that all observers can agree on. When the two ships come together it is B's clock that will register 2:15 while A's will register 3:15.

If, on the other hand, B had kept speeding onward at unchanging velocity, while A suddenly accelerated in order to chase after B and catch up with it, that acceleration would have made it possible for all observers to agree that it was A's clock that was slowed.

This effect, whereby all observers can agree that it is the accelerated object that has undergone the slowing of the time rate, is called time-dilatation, and it has an application to the Space Age.

The nearest star, Alpha Centauri, is just about 4¼ light-years away, a distance that comes to 25,000,000,000,000 (twenty-five trillion) miles (see Chapter 17). And since the velocity of light is the ultimate speed limit, it might seem that a trip from here to Alpha Centauri can never take less than 4¼ years.

In actual fact, a spaceship cannot reach velocities approaching that of light except by a long and gradual acceleration so that for a considerable period of time it travels at much below the velocity of light and should therefore take considerably longer than 4¼ years to reach Alpha Centauri.

But thanks to time-dilatation, this is not quite so. Suppose the ship accelerates at 1 *g* (an acceleration at which the crewmen will experience a feeling of weight directed toward the rear of the ship equal to that which they feel here on the earth). The combination of acceleration and rapid velocity introduces a slowing of time rate upon which all observers can agree.

To us on the earth, ten years might elapse while the space ship is *en route,* but to the crew on board ship, measuring the time lapse with clocks that move more and more slowly as their velocity increases, only 3½ years will pass before they reach Alpha Centauri.

As they continue to accelerate and their velocity approaches that of light more and more closely (though never quite matching the velocity of light) the time-dilatation effect becomes greater and greater. The ship can then cover perfectly amazing distances in what seems a comparatively short time to the crew.

Remember, however, that the time-dilatation is taking place only on the ship; not on the earth which continues at its accustomed velocity and which experiences time in the usual fashion. The time lapse which is short for the slow-time men on shipboard is therefore long for the fast-time men on the earth.

This can be shown, dramatically, by means of the following table which applies to a ship traveling out from the earth at a continual acceleration of 1 *g.*

Destination	Time Lapse on ship (years)	Time Lapse on earth (years)
Alpha Centauri	3.5	10
Vega	7	30
Pleiades	11	500
Center of the Milky Way	21	50,000
Magellanic Clouds	24	150,000
Andromeda Galaxy	28	2,000,000

So we can picture our astronauts visiting not only other

stars, but other galaxies, in a trip enduring a mere quarter century.

And this quarter century is not just a matter of clock measurement. It is not just the clock or other time-telling device that slows down on board an accelerating ship; *all* motion slows down. All atomic motion and, therefore, the rate of all chemical action, including that within an astronaut's body, slows down. Body chemistry proceeds at a slower rate. The mind thinks and experiences more slowly.

This means that under the effect of time-dilatation on the trip to the Andromeda Galaxy, the astronauts not only measure the time lapse as 28 years, but *experience* the time lapse as 28 years. What's more, their bodies age 28 years and no more, even though, in that same interval of time, two million years pass on earth.

Furthermore, this time-dilatation effect is something to which all observers can agree, so that if the astronauts were to return to earth, the earthmen of millions of years hence would have to admit that the astronauts had not aged more than a few decades.

This is the foundation of the "twin paradox." Suppose that a person heads out on a spaceship which accelerates steadily to high velocities, while his twin brother stays at home. The traveling brother gradually slows, comes to a halt, turns, speeds up and slows down again while returning to earth. Thanks to time-dilatation, he ages 10 years while his stay-at-home brother (along with everyone else on the earth) ages 40 years. When the traveler returns, he is 30 years younger than his twin brother.

Mind you, the traveler has *not* been rejuvenated, he has *not* grown younger. It is impossible for time to move backwards, and the traveler has simply aged less rapidly than he would if he had stayed put.

Nor has the traveler extended his life-span. If both he and his stay-at-home brother lived to a physiological age of 70, then the stay-at-home might die in the year (let us say) 2050, while the traveler survives to 2080. Still, though the traveler witnesses thirty years of events after his brother's death, he has not experienced thirty years more all told. While he was on his travels, he was experiencing only ten years while his stay-at-home brother was experiencing forty. Both would die with exactly seventy years of memories.

Even if the traveling brother had gone to Andromeda and back and had eventually died millions of years after his stay-at-home brother, both would have experienced just 70 years of life and memories.

Of course, there are experiences and experiences. There is something attractive about the thought of 70 years spent in moving out into space and back, touching the earth, let us say, at fifty-thousand-year intervals (earth-time). There is not only the experience of space-travel, but also the experience of what is virtually time-travel. Such a space-hopping astronaut would have the ability to witness mankind's future history vastly telescoped.

However, there is one drawback to this. Time-travel by way of the twin paradox is one-way only—toward the future. Once having set out along the road of time-dilatation, you cannot repent, you cannot return. The century of your birth will be gone forever and there will be no going back.

Chapter 19 The Birth and Death of the Universe

It isn't often that a scientist can make the front pages by giving up a theory, but Fred Hoyle, the English astronomer, managed to do so in the fall of 1965. He gave up on "continuous creation" because of objects fifty billion trillion miles away in space and ten billion years ago in time.

That's a long way to go from the here and now, but it was necessary to do so in order to settle the most grandiose clash of theories in all the history of science. Those theories involve nothing less than the birth (or non-birth) and death (or non-death) of the universe.

It began a half-century ago, when astronomers still knew very little about the universe outside our own Milky Way Galaxy—a lens-shaped conglomeration of a hundred thirty billion stars a hundred thousand light-years across. Here

This article appeared under the title of "Over the Edge of the Universe" in *Harper's Magazine*, March 1967. Copyright © 1967 by Harper's Magazine, Inc.

and there in the sky one could glimpse small patches of cloudy light which, some astronomers suspected even then, might be other conglomerations of stars, other galaxies. These might be millions of light-years away (with each light-year equal to a distance of nearly six trillion miles).

The light from these galaxies, or from any glowing heavenly object, can be gathered by means of telescopes, then spread out into a faint rainbow (or "spectrum") crossed by a number of dark lines. Each dark line is produced by a particular chemical element and has a particular place in the spectrum, *if* the light source is stationary with respect to ourselves. If the light source is receding from us, those lines would all be shifted toward the red end of the spectrum; the greater the velocity of recession, the greater the extent of this "red-shift." If the light source is approaching us, the lines would shift toward the opposite, or violet, end of the spectrum in a "violet-shift."

In 1912, the American astronomer, Vesto Melvin Slipher, began to collect light from the various galaxies in order to measure the nature and extent of the shift of the dark lines. He fully expected to find that roughly half would show a red-shift and half a violet-shift, that half were receding from us and half approaching us.

That proved not to be the case. To Slipher's surprise, only a few of the very nearest galaxies showed a violet-shift. The others all showed a red-shift. By 1917, he had found two galaxies that were approaching us and thirteen that were receding.

What's more, the size of the red-shift was unusually high. Individual stars within our own galaxy show red-shifts that indicate recessions of less than a hundred miles a second, but Slipher was detecting galactic recessions of up to four hundred miles a second, judging by the amount of the red-shift.

Others took up the task. Another American astronomer, Milton La Salle Humason, began exposing photographic film, night after night, to the light of very faint galaxies, allowing the feeble rays to accumulate to the point where a detectable spectrum would be imprinted upon the film. In this way he could measure the motions of particularly distant galaxies. *All* the faint galaxies showed a red-shift, with never an exception. And the fainter (and, presumably, more distant) they were, the greater the red-shift. By 1936, he was clocking velocities of recession of 25,000 miles per second, better than one-eighth the speed of light.

Already, in the late 1920s, the American astronomer, Edwin Powell Hubble, had generalized the matter, evolving what is now called "Hubble's Law." This states that the distant galaxies recede from us at a rate proportional to their distance from us.

According to present notions, this steadily increasing velocity of recession reaches a value equal to the velocity of light at a distance of about 12.5 billion light-years from ourselves. If a galaxy recedes from us at the velocity of light, the light it emits in our direction can never reach us. This means that nothing we do, no instrument we can possibly use, can detect that galaxy. We could not see its light, receive subatomic particles from it, or even detect its gravitational field.

The distance of 12.5 billion light-years represents, then, the edge of the "observable universe." Whether or not there is anything farther is of no moment, for nothing farther can impinge on us or affect us in any way.

There's our universe then; a gigantic sphere of space, pockmarked with galaxies, with ourselves at the center and with its edge 12.5 billion light-years away in every direction.

It seems odd, though, that we should happen to be at the center of the universe and that the galaxies should all be racing away from us. What is so special about *us?*

Nothing, of course. If there seems to be something special, it can only be an illusion.

Einstein's general theory of relativity, advanced in 1916, can be made to fit the view that the universe is expanding. As it expands, the galaxies within it find themselves scattered through a constantly enlarging volume of space. (The galaxies themselves, held together by gravitational force, do *not* expand.) Each one finds itself farther and farther from its neighbors as the universe expands.

In such an expanding universe, it would seem to an observer on *any* galaxy that all the other galaxies were receding from himself (except, possibly, for one or two very nearest ones that might be part of a common cluster of galaxies). What's more, it would seem to an observer on *any* galaxy in an expanding universe, that other galaxies receded at a rate proportional to distance.

It would seem, then, that the general appearance of the universe would remain the same regardless of the position in space from which it is viewed. This is called the "cosmological principle"—cosmology being the name given to that branch of science that studies the properties of the universe as a whole.

This expansion may simply be an intrinsic property of space, but in 1927, a Belgian astronomer, Georges Édouard Lemaître, advanced a physical explanation. The universe might be expanding because it was showing the effects of a colossal explosion that had taken place billions of years before. Originally, Lemaître suggested, all the matter of the universe had been collected into one solid, very dense mass of material—a "cosmic egg." This exploded in the vastest imaginable cataclysm and broke into pieces that eventually evolved into the present arrangement of galaxies. The galaxies are still separating from each other in consequence of that original explosion and thus create what seems to be an expanding universe.

Others have taken up this view since 1927 and have worked out its consequences in great detail. Perhaps the most vocal proponent of this "big bang" theory (as it is popularly called) is the Russian-American physicist, George Gamow.

The "big bang" theory envisages a universe that changes drastically with time. At first (about 10 to 15 billion years ago, astronomers now estimate) the universe was just a globe of superdense matter. Then it became an exploding mass of very hot fragments, very close together. With time, the fragments cooled off, spread apart, evolved into stars and galaxies, and continued to spread apart. Now the fragments are millions of light-years apart and as time goes on, they will become even farther apart.

The "big bang" theory, with its necessary view of a universe that changed with time, did not satisfy all astronomers. To three of them in England, Hermann Bondi, Thomas Gold, and Fred Hoyle, it seemed, in 1948, that the cosmological principle (by which the universe was assumed to appear generally the same to all observers) was incomplete if it referred only to observers at different places in space. They extended the notion to observers at different moments in time and called the result the "perfect cosmological principle." By this extended view, the universe as a whole did not change with time, but remained essentially the same in appearance throughout the eons.

But the universe *was* expanding, they admitted. The galaxies *were* drawing farther apart. To save their extended principle, Bondi, Gold, and Hoyle suggested that as the universe expanded, and as the galaxies moved farther apart, new matter was being continually created everywhere at an exceedingly slow rate, a rate so slow as to be indetectable to our most delicate instruments. By the time two galaxies had doubled the distance between themselves as a result of the expansion of space, enough matter had been created between them, however, even at this exceedingly slow rate, to conglomerate into a new galaxy.

In this way, although the universe expanded forever, the distances between neighboring galaxies remained always the same, for new galaxies formed within the sphere of the observable universe as fast as old ones moved outward beyond its limits. The appearance of the universe as a whole remained the same, then, through all the eternal past and into all the eternal future.

Each view—the "big bang" and "continuous creation"—has its separate beauty and each has its proponents, led by George Gamow and Fred Hoyle, respectively. Even among nonastronomers, emotional attachments were formed. Some people found themselves attracted to the colossal superspectacle of a huge "let there-be-light" explosion; while others found an austere glory in the thought of a universe without beginning and without ending, a universe that changed continually and yet remained always in the same place. But which theory, if either, is correct? Is there no way to choose between them?

Actually, the distinction between the two theories would be easy if only astronomers had a time machine. All one would have to do would be to get into the time machine and move ten billion years into the past (or into the future) and take a quick look at the universe. If it looks just about the same as today, then, the "big bang" cannnot be right and "continuous creation" will look good. If, on the other hand, the universe looks radically different from what it is today, then "continuous creation" can't be right and "big bang" will hold the field.

Oddly enough, astronomers *do* have a time machine, after a fashion.

Light (or any other form of radiation) cannot travel faster than 186,282 miles per second. This is fast, on the terrestrial scale, but it is a mere creep in the universe as a whole. Light from the most distant galaxies we can see takes a billion years or more to reach us. This means that what we see when we look at the very distant galaxies is the universe as it was a billion years ago or more.

All we have to decide, then, is whether what we see far, far away is essentially the same as what we see in our own neighborhood. If the very distant galaxies are just like the ones in our neighborhood, and show no change, then we can forget about the "big bang" (which postulates change). If the very distant galaxies are quite different from those in our own neighborhood so that there is a clear change with time, we can forget about "continuous creation" (which postulates no change).

But there is a catch. It is very difficult to see things at the billion light-year mark and beyond. All we can make out, at best, seem to be tiny patches of foggy light. If there are significant differences in the fine structure of those distant galaxies as compared to our own, we are almost bound to miss them. In order for a difference to be detectable across billions of light-years of space, it would have to be a huge and very general difference.

Through 1950, nothing of the sort had been detected. But a new technique had been devised in the meantime—a new kind of tool for peering into the ultimate depths of space.

It seems that in 1931, an American radio engineer, Karl Jansky, was engaged in the purely nonastronomic problem of countering the disruptive effects of static in radio communication. There was one source of static he could not at first pin down and which, he finally decided, had to come from outer space.

His announcement made no splash at the time. For one thing it seemed interesting but impractical. The radio waves from outer space were very short, and devices for detecting feeble beams of such radiation had not yet been developed. As it turned out, though, radar apparatus involved the detection of just such radiation and by the time World War II was over, the effort to make radar practical had resulted in new abilities to deal with shortwave radio from outer space. In this way "radio astronomy" was born and huge receiving devices ("radio telescopes") were turned on the heavens.

Radio waves were detected from the sun and from a few cloudy objects which seem to be the remnants of stars that had once exploded in a ferocious manner. Radio waves were even detected from the central core of our own galaxy, a core that is hidden from sight (as far as ordinary light is concerned) by the existence of vast clouds of light-absorbing dust between that core and ourselves—clouds that radio waves, however, could penetrate.

By 1950, over a thousand separate sources of radio-wave emissions had been marked out in the sky, but only a very few of them had actually been pinned down to something visible. The trouble was that even short radio waves are much longer than ordinary light waves; and the longer the waves, the fuzzier the "vision." Trying to find the exact source of a faint beam of radio waves was rather like trying to spot the exact source of a light beam viewed through frosted glass. All you see is a smear of radiation.

Nevertheless, a particularly powerful source of radio wave radiation (called "Cygnus A") had, with patience and perseverance, been boxed down to a very small area by 1951. Within that area, the German-American astronomer, Walter Baade, noted a peculiarly shaped galaxy. On closer study, the galaxy seemed to be not one, but two galaxies, the two being in collision. This seemed the source of that particular beam of radio waves; a pair of colliding galaxies 700,000,000 light-years away.

For the first time, it became clear that radio waves could be detected at enormous distances. Indeed, "radio galaxies" which emitted radio waves as powerfully as Cygnus A did could be easily detected at distances so vast that their light would be indetectable by even our most powerful ordinary telescopes.

Radio telescopes could reach out for unprecedented distances and, therefore, could reach back in time over an unprecedented number of eons.

This posed a very exciting possibility for astronomers. Suppose one made the assumption that all, or virtually all, the radio wave sources were far-distant galaxies which emitted radio waves in enormous concentration because they were colliding or exploding or undergoing some other huge catastrophe. To be sure only a very small percentage of galaxies were likely to be involved in such catastrophes but the universe contained many billions of galaxies so that it could easily contain a few thousand "radio galaxies." Those few thousand might be enough.

It seemed reasonable to suppose that the dimmer the radio wave source, the more distant the galaxy it represented. In that case, it was possible to count the number of such sources at various distances. If the "continuous creation" theory is correct, then the universe is always generally the same throughout time and there ought to be the same number of catastrophes taking place at all times. In that case, the number of radio sources for a given volume of space ought to remain at a steady value with increasing distance.

If, on the other hand, the "big bang" theory is correct, the youthful universe one detects at great distances must have been much hotter and more crowded than our present universe. In such a youthful universe catastrophes might reasonably be expected to be more common than in our own. Therefore, the number of radio sources for a given volume of space ought to increase with distance.

In the mid-1950s, the English astronomer, Martin Ryle, undertook a careful count of the radio sources and announced that the number of sources did, indeed, increase with distance as the "big bang" theory required.

Ryle's work was not completely convincing, however. It rested upon the detection and measurement of very faint radio sources, and even slight errors, which could easily have occurred, would have sufficed to wipe out completely the trend upon which Ryle had based his conclusion. The backers of "continuous creation" grimly clung, therefore, to their own view of the universe.

As radio-wave sources continued to be pinned down into narrower areas, several in particular attracted attention. The sources seemed to be so small that it was possible they might be individual stars rather than galaxies. If so, they would have to be quite close (individual stars cannot be made out at very great distances) and Ryle's assumption that all radio sources were distant galaxies would be upset; and with it his conclusion. "Continuous creation" would then gain a new lease on life.

Among the compact radio sources were several known as 3C48, 3C147, 3C196, 3C273, and 3C286. The "3C" is short for "Third Cambridge Catalog of Radio Stars," a listing compiled by Ryle and his group, while the remaining numbers represent the placing of the source on that list.

Every effort was made to detect the stars that might be giving rise to these 3C sources. In America, Allan Sandage was meticulously searching the suspected areas with the 200-inch telescope at Mount Palomar ready to pounce on any suspicious-looking star. In Australia, Cyril Hazard kept his radio telescope focused on 3C273 while the moon bore down in its direction. As the moon moved in front of 3C273, the radio-wave beam was cut off. At the instant of cutoff, the edge of the moon had obviously cut across the exact location of the source.

By 1960, the stars had been found. They were not new discoveries at all; they had been recorded on previous photographic sweeps of the sky but had always been taken to be nothing more than faint members of our own galaxy. A new painstaking investigation, spurred by their unusual radio-wave emission, now showed, however, that they were not ordinary stars after all. Faint clouds of matter seemed to hover about a couple of them, and 3C273 showed signs of a tiny jet of something or other emerging from it.

What's more, their spectra, when obtained by two American astronomers, Jesse L. Greenstein and the Dutch-born Maarten Schmidt, proved to be most peculiar. The few lines that were present were in locations that couldn't be identified with any known elements. It was a most puzzling mystery and was abandoned in frustration for a time.

In 1963, Schmidt returned to the spectrum of 3C273. Six lines were present and it suddenly occurred to him that four of these were spaced in such a way as to resemble a well-known series of lines that should be in a far different part of the spectrum. In order for these four lines to be in the place they were actually observed, they would have had to have undergone a red-shift of unprecedented size. Could that be? He turned to the other spectra. If he allowed very large red-shifts, he could identify every single one of the lines involved.

Within the next two or three years, the result of a concentrated search of the skies was to uncover about forty of these objects altogether. The spectra of more than half were obtained and all showed enormous red-shifts. One, in fact, is receding at a record velocity of 150,000 miles per second and is estimated to be about nine billion light-years away (fifty billion trillion miles).

And yet, if such red-shifts are allowed, then these apparent "stars" had to be very distant because, on the basis of the expanding universe, a large red-shift is always associated with huge distances. In fact, these queer objects had to be farther away than any other known bodies in the universe.

At such distances, what looked like stars certainly could not be stars. No ordinary star could possibly be seen at such huge distances. The objects were therefore called "quasi-stellar" ("star-like") radio sources, and quasi-stellar soon came to be shortened to "quasar."

The quasars are a rich source of puzzlement for astronomers. If the red-shift is interpreted in the light of the expanding universe and if the quasars are indeed billions of light-years away, then they have very unusual properties. To appear as bright as they do at such enormous distances, they must be glowing with the luminosity of ten to a hundred galaxies. And yet, there are many reasons for supposing that they are not very large in size. They may be only one to ten light-years in diameter rather than the hundred-thousand-light-year span of an ordinary galaxy.

What kind of a body can it be that has its substance

crowded into so tiny a fraction of a galactic volume and yet blazes with the light of dozens of galaxies? There are almost as many theories as there are astronomers—but as far as the fate of the "continuous creation" view of the universe is concerned, the theories don't matter. The mere fact that quasars exist might be enough.

The key point is that there are many quasars far away and no quasars within a billion light-years of ourselves. This means that there were many quasars in the long-gone youthful universe and none now. The number of quasars (which may be the source of all or almost all the radio-wave beams studied by Ryle) may increase with distance and, therefore, with the youthfulness of the universe. This means that we have detected one important change in the universe with advancing time—the number of quasars diminishes. That is enough to eliminate "continuous creation."

It is enough, that is, *if*, indeed, the quasars are far, far distant objects. The belief that they are rests on the assumption that the gigantic red-shifts they display are part of the expansion of the universe—but what if they aren't?

Suppose that quasars are small portions of nearby galaxies, hurled outward from the core of those galaxies by means of core-sized explosions. Examples of "exploding galaxies" have indeed been detected in recent years and astronomers are now carefully tracking down galaxies which for one reason or other —odd shapes, wisps of fogginess, signs of internal convulsion—look unusual. A few quasars have been detected not far from such "peculiar galaxies."

Is this coincidence? Do the quasars happen to be in the same line of sight as the peculiar galaxies? Or were they cast outward with monstrous velocities from those galaxies as a result of explosions involving millions of stars? If so, the quasars might not all be unusually far away from us after all. Some might be close, some far, and their distribution might not force us to give up the "continuous creation" theory after all.

This is possible, but there are arguments against it. Suppose that quasars are objects hurled out of galaxies with such force as to be traveling at large fractions of the speed of light. Some of them would indeed be hurled away from us and would show a gigantic red-shift that would be misleading if it were interpreted as representing a recession caused by the general expansion of the universe rather than by a special explosion of a galaxy.

A roughly equal number would, however, be hurled toward us and would be approaching us at large fractions of the speed of light. They would then show a gigantic violet-shift.

Then, too, some would be hurled neither toward us nor away from us, but more or less across our line of sight in a sideways direction. Such quasars would show only a small (if any) red-shift or violet-shift, but, considering how close they might be and how rapidly they might be moving, they would alter their positions in the sky by a slight but measurable amount over the couple of years they have been observed.

The fact is, however, that no quasars have been found that show a violet-shift, and none that alter position. Only red-shifts have been observed, gigantic red-shifts. To suppose that comparatively nearby explosions have cast out quasars in such a way as to produce red-shifts only is to ask too much of coincidence.

So the weight of the evidence is in favor of the great distance of the quasars and of the elimination of the "continuous creation" theory—and Fred Hoyle gave up.

The elimination of "continuous creation" does not necessarily mean the establishment of the "big bang." Suppose there is some third possibility that is as yet unsuggested. To strengthen the "big bang" theory against the general field of unsuggested possibilities, it would be nice to consider some phenomenon that the "big bang" theory would predict, some phenomenon that could then actually be observed.

Suppose, for instance, that the universe *did* begin as an incredibly dense cosmic egg that exploded. At the moment of explosion, it must have been tremendously hot—possibly as hot as 10 billion degrees Centigrade (equivalent to 18 billion degrees Fahrenheit).

If so, then if our instruments could penetrate far enough, to nearly the very edge of the observable universe, they might reach far enough back in time to catch a whiff of the radiation that accompanied the "big bang."

At temperatures of billions of degrees, the radiation would

be in the form of very energetic X rays. However, the expanding universe would be carrying that source of X rays away from us at nearly the speed of light. This incredible speed of recession would have the effect of vastly weakening the energy of the radiation; weakening it to the point where it would reach us in the form of radio waves with a certain group of properties. Through the 1960s, estimates of what those properties might be were advanced.

Then, early in 1966, a weak background of radio-wave radiation was detected in the skies; radiation that would just fit the type to be expected of the "big bang." This has been verified and it looks very much as though we have not only eliminated "continuous creation" but have actually detected the "big bang."

If so, then we have lost something. In facing our own individual deaths, it was possible after all, even for those who lacked faith in an afterlife, to find consolation. Life itself would still go on. In a "continuous creation" universe, it would even be possible to conceive of mankind as moving, when necessary, from an old galaxy to a young one and as existing eventually through all infinity and for all eternity. It is a colossal, godlike vision, that might almost make individual death a matter of no consequence.

In the "big bang" scheme of things, however, our particular universe has a beginning—and an ending, too. Either it spreads out ever more thinly while *all* the galaxies grow old and the individual stars die, one by one, or it reaches some maximum extent and then begins to collapse once more, returning after many eons to a momentary existence as a cosmic egg.

In either case, mankind, as we know it, must cease to exist, and the dream of godhood must end. Death has now been rediscovered and Homo sapiens, as a species, like men as individuals, must learn to face the inevitable end.

—Or, if the universe oscillates, and if the cosmic egg is reformed every hundred billion years or so, to explode once more; then, perhaps, in each of an infinite number of successive universes, a manlike intelligence (or a vast number of them) arises to wonder about the beginning and end of it all.

PART II

CONCERNING THE MORE OR LESS UNKNOWN

A
OTHER LIFE

Chapter 20   A Science in Search of a Subject

I suppose we have all heard the wry comment about the miracle drugs being produced in such profusion that some are for diseases which have not yet been discovered. This idea of a cure without a disease is analogous to the state of the new science of "exobiology"—a field of study with nothing to study.

The word exobiology was coined by the Nobel Prize-winning American biologist, Joshua Lederberg. It means "outside-biology," the study of life forms outside the earth.

What life forms outside the earth?

There's the catch. We know of none, yet we suspect that some exist. Out in space, there must be stars like our sun around which circle planets like our earth upon which are life forms like—what? Like ourselves? Nearly like ourselves? Like nothing we have ever dreamed of?

We don't know.

Here in the solar system there may be forms of life present on Mars—perhaps even on the moon. If so, what kind?

We don't know.

But speculation is free, and if we lack actual objects to study, there is no lack of concepts to consider in the quiet of the mind. In that sense, Lederberg is an exobiologist; so are such astronomers as William M. Sinton at Lowell Observatory, Stephen H. Dole at Rand Corporation, Carl Sagan at Harvard Observatory, and chemists like Harold C. Urey at the University of California.

Dole, for instance, in his book *Habitable Planets for Man,* comes to the conclusion (see Chapter 22) that in our galaxy alone there are likely to be some 640,000,000 earthlike, life-bearing planets. (And there are many billions of other galaxies.)

Sagan goes even further. He thinks it a reasonable guess that there may be as many as 1,000,000 planets in our

This article appeared in *The New York Times Magazine,* May 23, 1965.

galaxy that not only bear life, but bear intelligent life and advanced civilizations. He has even wondered if perhaps intelligent life forms from other worlds visited earth in the distant past, and cites ancient Babylonian myths to the effect that civilization was founded by nonhuman creatures of great learning.

But how do you speculate when you have nothing to go on, when there is not even the tiniest fragment of outside life to serve as a guide?

The answer to that is we *do* have something to go on. We know of one planet that is thoroughly infested with life— our own. Although one might suppose that it would be risky to draw conclusions about life in the universe generally from life on our single, insignificant planet, that it would be cheaply self-centered to do so, there are, in fact, reasonable lines of argument that give us a strong justification for doing just that.

In the first place, earth is not an odd or unusual planet, chemically speaking. Astronomers, in their study of the composition of stars and of the material between the stars (based on the nature of the light given off or absorbed) have worked out definite notions as to the relative abundance of the different chemical elements in the universe.

The two most abundant elements are the light gases, hydrogen and helium. The earth's gravity was too weak and its temperature too high during the process of planetary formation for those gases to have been retained. Certain other gases such as neon and argon were also lost, but except for these, the structure of the earth is similar in nature and proportion to that of the universe generally.

The earth is therefore a normal and typical planet—not one built up of rare elements that through some freak of nature made it, and it alone, suitable for life. In fact, if we find a planet anywhere in the universe with a mass and temperature roughly like that of the earth, we can be almost sure it will be very like the earth structurally and chemically.

Given a planet like earth, then, what kind of life may we expect to find on it? To answer that, we should first see what kind of life is possible.

A SCIENCE IN SEARCH OF A SUBJECT

On all of earth, there is only one basic form of life. All earthly life, from the simplest virus to the largest whale and redwood tree, is based on proteins and nucleic acids (see Chapter 6). All make use of the same vitamins, the same types of chemical changes, the same methods of liberating and utilizing energy. All of life follows a single pathway, however much particular species may seem to vary in detail.

Furthermore, earthly life, which began in the sea, is made up precisely of those elements that are, and were, common in the sea. There are no "mystery ingredients," no rare and magical items that were included only through a stroke of great good fortune.

Another planet, with the mass and temperature of earth, might also be expected to have watery oceans, with the same type of dissolved salts. It should, therefore, develop life based on the same chemical elements as ours. Does it follow, then (once we have gone this far), that it must also move along the same general pathway that life follows on earth?

Here we must hesitate. Elements can be put together in a vast number of different ways. Suppose that in the early days of our own planet, when life was first forming the primordial ocean, a thousand different schemes of life set sail. Let us further suppose that one particular scheme won out over the rest, perhaps through the sheerest chance; the survival of that one scheme could now give us the false impression that it is the inevitable and only possible scheme.

That may be so, of course, but what evidence we have points in the other direction. In the 1950s and 1960s chemists have tried to duplicate the chemical conditions that existed on the primordial earth, and have observed what complex molecules would evolve spontaneously from the simple substances that then existed (see Chapter 9).

The compounds that formed are the familiar compounds that make up our body—amino acids out of which our proteins are built, nucleotides out of which our nucleic acids are built, porphyrin rings out of which chlorophyll and hemoglobin are built.

All of the substances formed out of systems imitating the primordial ocean are on the broad highway leading to our particular kind of life. There is no sign yet of any turn-off, no appearance of any side street. One may yet appear in the future, but experiment after experiment is decreasing that probability.

On any planet like ours, then, the chemical basis of life may quite likely be the same as on earth. We have no reason as yet to expect otherwise. Furthermore, the general trend of evolution ought to be the same. The pressures of natural selection tend to fill all possible regions of a planet with organisms adapted to those regions. On earth, after the development of life in the sea, there was a gradual invasion of fresh water by organisms adapted to conserve salt, an invasion of dry land by organisms adapted to conserve water, and an invasion of the air by organisms adapted to flight.

All this should happen on another planet, too, and there should be a certain limit on novelty. On any earthlike planet a flying creature can be no more than a certain size if the air is to support it; a sea creature must be either stream-lined or slow-moving, and so on.

It is quite reasonable, then, to expect other-worldly life to develop recognizable features based on general utility. It ought to have left-right symmetry. It ought to have a distinct head in which the brain and sense organs are concentrated. Among the sense organs there ought certainly to be those which can sense light, like our eyes. The more active forms ought to eat the plantlike forms, and it is very likely they will breathe oxygen, or absorb it in some fashion.

In short, life on any earthlike planet ought not, perhaps, be completely alien. Undoubtedly, it will differ drastically and unpredictably in detail. (Who could have predicted the shape of the duckbill platypus before Australia was discovered, or those of deep-sea fish before they were actually seen?)

Life can vary in so many small details and in so many directions. Even though the chemistry is the same and the general structural plan is similar, the possible variations on a theme are so great in number that it is extremely unlikely that, through sheer chance, the same variations would occur on another planet as on earth. It would be entirely too much of a coincidence for an extraterrestrial creature to look like a man; even a vague resemblance might be too much to expect. Nevertheless, factors we hold in common would make it possible for us to accept such other-worldly life, if not as brothers, then at least as second cousins.

But, unfortunately, there are no truly earthlike planets within reach. Inside our solar system, Venus is earthlike in mass but far too hot for anything approaching our kind of life. Mars, on the other hand, is almost earthlike in temperature (a bit on the cold side) but is only one-tenth as massive as the earth and therefore has retained very little atmosphere. In particular, it possesses no oxygen and almost no water.

But is oxygen necessary for life? The oxygen in our own atmosphere is very likely there only because it is produced by green plants (see Chapter 13). Before green plants evolved there was probably no oxygen in the air, and life must have begun without it. Even today there are still certain bacterial forms which do not require oxygen to survive. To some, oxygen is actually poisonous. These may be the remnants of life surviving from the no-oxygen period of earth's early history.

We have no evidence that there ever was no-oxygen life more advanced than bacteria, but we can't be sure. Still, it would be best to assume that life on Mars, since it must be on the no-oxygen level, can only be very simple.

Through the early 1960s there had been increasing hope that just such simple plant forms might indeed exist on Mars. There are green patches on Mars which vary with the season of the year, as though vegetation were sometimes spreading, sometimes retreating. Sinton had studied reflected light from Mars and deduced the presence of chemicals resembling those found in plants. Certain simple forms of plant life here on earth have been grown under Martian conditions—intense cold, little water, no oxygen—and survived. In fact simple forms of life, such as bacteria and fungi, have even survived exposure to conditions similar to the even more hostile atmosphere of Jupiter, which is loaded with ordinarily poisonous methane and ammonia.

Unfortunately the signs of life on Mars are uncertain and have been somewhat discredited. Sinton has found that the reflected light from Mars could be interpreted in ways which did *not* involve plant life. Sagan has evolved a theory that would explain the spreading and retreating green patches without having to postulate forms of life. Worst of all, the Mars-probe, Mariner IV, flew by Mars in July 1965 and took photographs which showed the surface of that planet to be pockmarked with craters. The existence of such craters seemed to indicate the absence of erosion and therefore a longtime absence of water—something that lowers the chances of life ever having developed there.

Still, all hope is not gone. Some astronomers, including Sagan, still argue the possibility of Martian life; and while the chance of it is admittedly not high to even the most optimistic proponents, one of the most fascinating prospects of Martian exploration remains the chance of studying outer life. If such life is present on Mars, in even the simplest form, the science of exobiology will have taken a giant step forward.

Suppose the basic chemical structure of Martian life (assuming that any exists) is the same as ours, that life forms are built up of proteins and nucleic acids constructed out of the same simple building blocks as ours are. In that case the supposition that all life is basically one, on any planet even remotely resembling the earth, is immeasurably strengthened.

On the other hand, if Martian life forms are basically different in chemistry, that would be better still. For the first time, scientists would have a life scheme to study other than our own. The knowledge they might then gain of the general nature of life (the factors held in common by two basically different life schemes) could be of incalculable importance.

Scientists are not willing, therefore, to wait until men actually land on Mars to determine whether life exists there. Instruments are being developed that can be landed on Mars to check, automatically, for the presence of life. (This is the field of "applied exobiology.") These instruments are devised to eject sticky strings or other gadgets that will pick up Martian dust and particles and retrieve them. The dust and particles, possibly carrying living cells, will be immersed in liquids containing salts and nutrients of the type that would support earthly life, and instruments will then record and transmit to earth data on any changes in the cloudiness of the liquid or its acidity. Or they will record the development of carbon dioxide or the presence of specific reactions that could only be brought about by enzymes.

Any such changes would be strong evidence not only of the presence of life, but also of the presence of life based on the same chemical principles as those of earth.

But what if *no* changes are detected? Has Mars, then, no life? Or has the instrument happened to land in a barren area? Or do Martian life forms refuse to live and grow on the chemicals we send them? We could not be sure. We would then have to wait until we actually landed on Mars.

Perhaps the moon may give us a hint. We will be on the moon in a few years and, although it seems to have no air and no water, it may have life just the same. Traces of water and air may linger just under the surface or in the recesses of craters, and simple forms of life may exist there. If moon life is basically different from that on earth, the result will be as satisfactory as any obtained from a trip to Mars.

On the other hand, if moon life is based on earthly chemistry, we cannot be perfectly sure of its significance. Earthly objects have already landed there and, despite our efforts to sterilize them, they may have contaminated its surface.

Worse still, some astronomers believe that in the past, when earth and moon were closer to each other and meteoric bombardment was fiercer, material from one world may have splashed on the other. Urey recently speculated that enough earthly water splashed on the moon to give it a few short-lived lakes. In that case, the moon may have been contaminated by earthly life eons before the space program started, and we might have to wait for Mars to get a clear look at true exobiology.

Yet, despite all such speculation, we must go back to the initial statement that exobiology really lacks a subject to study. So far, we have only speculation; attractive speculation, it must be admitted, but nothing substantial.

Many biologists (notably the important Harvard zoologist, George Gaylord Simpson—himself a science fiction reader and by no means stodgy or unimaginative—and Theodosius Dobzhansky of Rockefeller University, likewise a man of brilliance and mental daring) are rather out of patience with what they consider overenthusiasm for a science that as yet is empty of real content.

Undoubtedly, then, it will pay exobiologists to proceed slowly, step by step.

Step 1. They must ground themselves firmly on the one type of life we know, that of the earth.

Step 2. They must see how their modest conclusions, based on what evidence can be gathered on earth, stack up against the reality of the moon and Mars when those objects are reached by man or by the proper instruments.

Step 3. —Well, let's wait for Step 2 before going on to Step 3.

Chapter 21   We, the In-Betweens

Here on earth, life has developed in many directions and has fitted snugly into a tremendous variety of environments, taking on forms that could scarcely be invented by the wildest imagination if they were not already known to exist.

Yet all the variations and modifications that exist on earth are in some ways only superficial. For all its wonderful differences, life on earth is merely a long-playing, imaginative variation on a single chemical theme (see Chapter 20), and life on any earthlike planet may prove to be a series of variations on that same perpetual theme.

Perhaps this should not be too surprising. As far as we understand life, it must consist of molecules that are large enough and complex enough to live up to the flexible and all-but-infinite requirements of living tissue. They must be stable enough, in spite of their complexity, to retain their structure under some conditions, and unstable enough to change kaleidoscopically under other conditions. Such large and complex, stable but unstable molecules are not easy to come by. In living things on earth, the most important mole-

First published in *Mademoiselle*, May 1961. Copyright © 1961 by Street & Smith Publications, Inc.

cules of this type are the proteins, and, as far as we know, nothing else will substitute for them.

Furthermore, the changes these proteins undergo in the business of living can only take place against a watery background. Life began in the oceans, and even the various forms of land life are still from 50 to 80 percent water.

The chemical theme, then, upon which life plays its variations, here and possibly on all earth-type planets, is protein-in-water (with the protein structure supervised by a complex nucleic acid system). If we are ever to meet up with creatures from an earth-type planet, we may not be able to predict beforehand whether they will sport wings, tentacles, green skin, ten feet, domed heads or bifurcated tails. But we can predict that, whatever their shape, they will very likely be protein-in-water under the supervision of nucleic acids.

But what about life on planets that are *not* like the earth? What about planets so close to their sun that their surfaces are hot enough to melt lead? What about planets so far from their sun that water is eternally frozen to rock-hard ice? Are such worlds doomed to be perpetually barren? It would seem so, certainly, if all life were only protein-in-water.

But is all life only that? Can we be sure that life cannot be based on other themes?

Suppose, for instance, that on a world on which liquid water does not and never did exist, thanks to the eternally frigid temperature, there was a substance which at those low temperatures could take the place of water. Actually, there is such a substance, and it is called ammonia.

Everyone is probably familiar with the bottled "ammonia" that looks like water but has a pungent smell. This is not actually ammonia, but only a solution of ammonia in water.

Ammonia itself is a gas at ordinary temperatures: pungent, eye-watering, poisonous. Under conditions on earth it doesn't become a liquid until it is cooled to thirty degrees below zero Fahrenheit. It doesn't freeze until a temperature of one hundred degrees below zero Fahrenheit is reached. The exact extent of its liquid range changes with the pressure of a planet's atmosphere, but under any conditions it would remain a liquid fifty degrees or more below the point at which water freezes.

Now the cold worlds of our own solar system, such as Jupiter and Saturn, have thick atmospheres that are mainly hydrogen and helium, but which contain a strong admixture of ammonia and methane. Perhaps some of the larger satellites of these planets also have such atmospheres. In fact, there is good reason to think that any large cold planet would have an atmosphere of this very sort.

It is conceivable, then, that such planets, even with all water frozen into solid ice, might have oceans of liquid ammonia in which life might develop in a completely alien manner.

Actually, the chemical behavior of ammonia strongly resembles that of water. Chemists have worked out and demonstrated a chemistry of substances dissolved in ammonia that is analogous to the ordinary chemistry of substances dissolved in water; so the theme of protein-in-ammonia is fascinatingly possible under conditions where the temperature is too cold for protein-in-water.

A life chemistry based on this new theme would be bound to differ radically from anything we know. Our proteins, sufficiently active to participate in life processes at our customary temperatures, become sluggish at the temperatures of liquid ammonia—too inert, perhaps, to support the quick-changing complexities of life. Still, there are many chemical structures that are too active, too unstable even, to exist for more than a split second at the temperature of water. These structures may become just stable enough at lower temperatures to provide a practical basis for life.

Again, organisms on earth eat food that contains complex molecules rich in carbon and hydrogen atoms. (Plants don't eat such food, but manufacture the complex molecules by using the energy of sunlight.) The hydrogen atoms are combined with the oxygen absorbed from the atmosphere, and the energy that is released supports life.

But on cold planets there is no oxygen in the atmosphere. There is, instead, hydrogen. Perhaps the food of the ammonia creatures would be complex molecules rich in carbon and oxygen atoms—molecules of types that might be too unstable to exist at the earth's high temperature. The oxygen atoms of such food could be combined with hydrogen absorbed from the atmosphere. Energy would, after all, be obtained as easily in this "reverse" manner as in our "right-side-up" way.

Even if a planet were too cold for ammonia to remain liquid (and the outermost planets in our solar system—Uranus and Neptune, for instance—are indeed too cold for that), all hope for other forms of life is not lost. There is methane, which, on earth, is the principal constituent of "natural gas." When burned, this is used to cook food and heat houses. Methane is harder to liquefy even than ammonia; down about the range of three hundred degrees below zero Fahrenheit it becomes liquid.

Methane, however, is a substance completely different in chemical properties from either ammonia or water. Ordinary proteins would not mix well with it, as they would with ammonia or water. Fatty substances would do so, nevertheless, and perhaps on very cold planets, complex fatty molecules will take the place of protein. Complex fatty molecules do in fact exist, even in organisms on the earth, and some of them are nearly as complex as proteins; so the notion of a fat-in-methane life theme is not entirely inconceivable.

What about hot planets, close to a sun? Such hot planets would be small and without an atmosphere of the usual type. Unappetizing gases, such as sulfur vapor and mercury vapor, might cling to it in small quantities. Certainly there would be no water. If any had existed at the beginning, it would have boiled away eons ago.

Perhaps life would develop in substances that are liquid at high temperatures. Sulfur (which is rather like oxygen chemically) is liquid between temperatures of 235 and 800 degrees Fahrenheit. Could there be sulfur-based life?

If there is, it could scarcely be based on ordinary protein, which would be highly unstable at such elevated temperatures. Ordinary protein and all the complex molecules in living tissue, including the molecules of nucleic acid that run the whole show, are made up for the most part of carbon and hydrogen atoms, with some oxygen, nitrogen, sulfur, and phosphorus atoms thrown in as minority constituents. Our ordinary molecules are, in other words, derivatives of hydrocarbons.

During World War II, however, as a result of research on the atomic bomb, chemists found that the hydrogen atoms in these molecules could be replaced by fluorine atoms (fluorine is a poisonous and exceedingly corrosive gas). The resultant fluorocarbon molecules have some properties like those of hydrocarbons but are much more stable. Complex chemicals built up out of derivatives of fluorocarbons would seem to be too stable to make up a living creature, but at the temperature of liquid sulfur they might be unstable enough. (It is difficult to judge from simple molecules of a particular type what the properties of complex varieties of the same type might be. For instance, a man-made molecule such as nylon has basic similarities to the atom combinations in proteins. If nylon, so stable and inert, were the only compound of its type you could study, who would ever predict the possible existence of the complex, unstable proteins with their reactivity and versatility?)

Another type of molecule that could conceivably be built up into complex structures able to survive elevated temperatures is the silicones. These are made up, essentially, of chains of silicon and oxygen atoms, as are the rocks of our planet. Attached to these chains, and lending them versatility, however, are hydrocarbon groups (or possibly fluorocarbon groups at high temperatures).

Such silicones have been developed in the laboratory in the last several decades here on earth. Solid silicones serve, among other things, as a kind of artificial rubber, while liquid silicones have been used as hydraulic fluids. Can we picture life forms on hot planets with rubbery tissues and hydraulic-fluid blood streams, living in puddles of liquid sulfur?

On hot planets it might not be necessary for a creature to use chemical reactions as a source of energy. With a blazing sun at least ten times as large and as bright as the one we see here on earth, such creatures, fluorocarbon or silicone, might be able to soak up energy directly out of the sunlight furnace.

Can we hope to come across such creatures in the future? After all, even if we are never able to reach the stars, we should, by our grandchildren's time, be able to reach the planets of our own solar system. And there, with the exception of Mars and its possible simple plants, we will be investigating worlds that are not in the least earthlike. What will we find on a hot planet like Mercury? Nothing but dead rock and steaming sulfur? What will we find on a cold world like Titan, Saturn's largest satellite? Nothing but hard ice and frigid methane winds?

We can't be entirely sure.

We are already undergoing a radical broadening of thought in beginning to accept the fact that we may not be the only world of living creatures in the universe; not even, perhaps, the only living intelligences. Will we someday have to undergo another broadening of thought and accept ourselves as an example of only one of the possible chemical themes of life?

If so, is it even vaguely possible that in the long run we will find ourselves studying, with fascination, the absolutely alien life chemistry of the fluorocarbon or silicone Hots, and the ammonia or methane Colds, with ourselves merely examples of the protein-in-water In-Betweens?

Why not? In science, as in everything human, it is the chance of the unexpected that lends spice to endeavor.

Chapter 22   Is Anyone There?

*Sit, Jessica. Look how the floor of heaven
Is thick inlaid with patines of bright gold:
There's not the smallest orb which thou behold'st
But in his motion like an angel sings,
Still quiring to the young-eyed cherubins.
Such harmony is in immortal souls;
But whilst this muddy vesture of decay
Doth grossly close it in, we cannot hear it.*

Thus spoke Lorenzo in Shakespeare's *The Merchant of Venice,*

This article appeared under the title of "Hello CTA-21—Is Anyone There?" in *The New York Times Magazine,* November 29, 1964.

as he yearned to hear the music of the spheres, and knew he could not.

Since Shakespeare's time, man has overcome part of the handicap of his "muddy vesture of decay" by means of new instruments: telescopes, spectroscopes, cameras, and microwave amplifiers. Now we can pick up the singing of the orbs in a very literal way, for the universe broadcasts radio waves. Translated into sound, they seem to make a coarse and rasping static, but to the ravished ears of astronomers, the crackling is angelic indeed.

From some invisible spots in the heavens come waves not quite like those in others. Two such spots were first observed in 1960 and later included in a listing of heavenly radio sources drawn up by the California Institute of Technology. From their numbers on that list, the sources in question are called CTA-21 and CTA-102. In 1963, a team of Anglo-American astronomers pointed out these sources as worthy of special study, and in October 1964, a leading Soviet astronomer, Nikolai S. Kardashev, supplied some of that study.

He came to the conclusion that the natural phenomena of the inanimate universe might not be responsible for the broadcasts from CTA-21 and CTA-102. Instead, he suggested, it was just conceivable that we might be observing radio beacons sent out by intelligent beings of high technological proficiency.

Should this be dismissed at once as fantasy? Not at all! Highly unlikely, of course (as Kardashev would himself admit), but not fantasy. Since World War II, astronomers have grown more and more convinced that somewhere out in the infinite depths of space are, indeed, other intelligences. This has come about chiefly because of changing theories concerning the origin of the solar system, and of life.

There are two general kinds of theories about the origin of the solar system: catastrophic and evolutionary. According to the first, as two stars pass close to each other, huge tides of matter are pulled out of each star and these condense to form planets. According to the second, a star is formed out of a huge cloud of swirling dust and gas, and out of the material at the edges of this cloud, planets are automatically formed as a star takes shape at the center.

During the first half of the 20th century, the catastrophic theory was generally accepted. As the nature of the interior of stars came to be better understood, however, astronomers threw it out. Material pulled from the sun by an approaching star could not condense to form planets. The extruded material would be too hot.

In 1944, a German astronomer, Carl F. von Weizsäcker, put forth a new version of the evolutionary theory which met with wide approval. Astronomers may argue over just how to modify it to meet various difficulties but virtually all agree, now, that some version of the evolutionary theory is the most useful way of looking at the matter.

This has an important bearing on the question of whether other intelligent creatures exist. If planets originate in catastrophes, then there can be very few of them in the universe, for stars virtually never come close to each other.

If, however, planets originate as part of the natural evolutionary changes undergone in the formation of a star, then they must be exceedingly common. Practically every star ought to have a train of planets—and this is what astronomers now believe.

How many of these planets are sufficiently like the earth, however, to qualify as possible abodes of life as we know it? Dr. Stephen H. Dole of the Rand Corporation has tried to answer that question on the basis of present knowledge.

In our own galaxy, the Milky Way, he points out, there exist an estimated 135 billion stars. Of these, however, only stars of a certain size range would make suitable suns for planets like ours. Particular planets would have to be of a certain size, a certain distance from the star, turn with a certain period of rotation, and so on, before they could be truly "earthlike."

Taking all reasonable considerations into account, Dr. Dole concludes that there are some 640 million earthlike planets scattered here and there in our own galaxy.

If these earthlike planets are distributed evenly throughout our galaxy, then the nearest would be 27 light-years away (a distance equal to 150 million million miles). Within

100 light-years of earth, in all directions, there might be as many as 50 earthlike planets.

Could these planets bear life? Right now the conclusion is: Yes, almost certainly. Recent experiments seem to show that life is no rare accident arising out of an unusual combination of chemicals, but that it would tend to originate anywhere where the conditions were similar to those on the primordial earth (see Chapters 20 and 21).

But how many of these planets would bear *intelligent* life?

Ah, there science is still stumped completely. There is no way of telling. Life on earth existed for some two or three billion years before an intelligent species developed. And might this not have been but a rare and lucky accident? Might it not be far more likely that life might have continued throughout the entire lifetime of a planet without happening to develop intelligence?

We don't know the answer to that (and Dr. Dole ventures no guesses), but even if intelligence rose only on one out of a million life-bearing planets, there might still be nearly a thousand intelligent species scattered throughout the galaxy. And if this is so, then the activities of some of them may give them away, if we listen carefully enough and subtly enough—especially if, for some reason, they are trying to make themselves heard. It is not likely that we will hear anything by listening to the universe; but it is not impossible, either.

If we wanted to send a message to some life form on a planet circling another star, or to receive a message from it, some signal that could reach across vast gulfs of space is necessary. We ourselves receive three types of such signals from outer space. They are (1) gravitational effects, (2) streams of subatomic particles, and (3) electromagnetic radiation.

Of these three, gravitational force reaches us most strongly from the sun and the moon. Our path about the sun is in response to its giant pull, and the ocean tides rise in response to the moon's. The weaker pulls of Venus and Mars can be detected in small variations in the moon's motion.

However, gravitational force is the weakest force in nature, and it reaches us from other stars with an intensity so weak that there is no practical way of detecting it. Nor could we send out a useful gravitational beam even if it were a stronger force than it is, since we know of no practical way of turning gravity on and off in order to send out a gravitation dot-dash code, for instance.

Streams of subatomic particles (objects far smaller even than atoms) reach us in the form of protons and electrons from the sun and in the form of cosmic rays (very high energy protons and still more massive electrically charged particles) from farther out in space. We can produce streams of such particles easily enough and turn them on and off, too, but only in small quantities.

Even if we could produce them in mighty streams with a force that would squirt them from star to star, we couldn't send them outward in a smoothly aimed line. The paths of electrically charged particles would curve and veer whenever they passed through the magnetic fields that are so thickly strewn through space. Furthermore, they, together with most uncharged particles, would be absorbed and changed by the atmosphere that would undoubtedly surround an earthlike planet.

One type of subatomic particle, the neutrino, suffers from none of these disadvantages. It could be made to travel in a straight line from star to star, and would be affected by neither gravity, magnetic fields, nor atmospheres. Unfortunately, this particle is nearly impossible to detect.

That leaves electromagnetic radiation, of which two types penetrate our atmosphere. One is ordinary light and the other high-frequency radio waves of a type usually termed "microwaves." Both are easy to produce, easy to detect, are not affected by magnetic fields or atmospheres, and, in short, are nearly ideal for the purpose.

Of the two, light might seem to be the first choice. You can easily imagine a huge searchlight sending out flashes in Morse code toward the stars. There are some basic difficulties to this, however.

First, there are a great many light sources in the galaxy, considering its billions of stars, so that one dim signal would be lost among them. In particular, the light originating on some far-off planet would be bound to be blotted out by the superior light of the planet's own sun. To be sure, one could argue around this. Suppose the beacon light were that of a gigantic laser (see Chapter 11). The characteristic light of a laser could be differentiated from that of stars and, indeed, the mere existence of laser light might well be considered a sign of intelligence at the other end. An even more daring suggestion is that a sufficiently advanced civilization might learn to use the stars themselves as beacons. Thus, some of the quasars (see Chapter 19) vary in light intensity with time. Could some superbeings be using them to send a kind of Morse code? Not at all likely, I hasten to emphasize, but very interesting to think about.

Another difficulty with light, however, is that it cannot penetrate great thicknesses of dust and our neighborhood of the galaxy is pretty dusty. We cannot see the glorious burst of light of the billions of stars in the core of our galaxy, for instance. Dust clouds block it off.

That leaves microwaves. These penetrate dust clouds nicely, and we can detect microwaves coming from the core of our galaxy without trouble.

The sources of microwaves in the sky ("radio sources," some of which are visible by the light they also emit, but most of which have not yet been associated with visible objects) are far fewer than are the sources of light. That makes an oddly behaving radio source far easier to spot. Furthermore, a strong radio source on a planet is not likely to be blotted out by the sun of that planet, for few stars are strong microwave emitters.

It is easy to measure the length of the individual waves of the microwave beam arriving from outer space. From most radio sources, this "wavelength" is a matter of feet and yards. However, for purposes of communication, it would be better to use shorter microwaves. It is suggested that wavelengths of 3 to 6 inches would be ideal. Such waves would be least likely to undergo distortion or interference on long voyages or to be drowned out by natural sources of microwaves.

That is why the emissions of CTA-21 and CTA-102 rouse such interest. The microwaves received from these sources are chiefly in the 4- to 20-inch range, with a peak at about

12 inches or so. This isn't quite ideal, but it is fairly close, much closer than is true for other sources. Furthermore, as best astronomers can tell, those microwaves arise from a tiny "point-source" in the heavens, as though they were originating from planets. In the case of the usual radio source, the origin is more extended, indicating the source to be a large volume of gas.

If, indeed, the microwave emissions of CTA-21 and CTA-102 are the product of intelligent life, then they must represent civilizations far more advanced than our own.

Right now, mankind on earth is producing power at the rate of 4 billion kilowatts. Even if all of this were poured into a microwave beacon and sent out into space it would not suffice. The beacon would spread out and grow dilute, even though it were made as coherent as possible, and by the time even the nearest intelligent beings had been reached, it would have grown too feeble to detect. To produce beacons strong enough to detect would require a civilization capable of wielding far more energy than we do.

Mankind's energy output is growing at the rate of 3 to 4 percent a year. If nothing happens to interfere with this, then in a matter of 3,200 years we will be producing energy at the rate of the sun and we could then announce our own existence with beams that will stretch through the length and breadth of our galaxy. And if we can detect the beams of other life forms *now*, then those life forms must be at least several thousand years ahead of us in technology.

To be sure, one ought not to take too seriously the specific cases of CTA-21 and CTA-102. They are enormously distant objects that are probably quasars, and no doubt their microwave radiation can be explained without having to assume intelligences out there.

Still, suppose some intelligence on some fairly nearby star *is* trying to reach us. Or suppose we try to reach them. What is there to say in the announcements being sent out or received? We can't really use Morse code or expect any foreign intelligence to speak English. We must find something universally understandable. We could assume, for instance, that the people of any supercivilization would understand mathematics and that whatever mathematical statements are true here are also true there.

For instance, suppose we sent out two pulses of microwaves, followed by two more and then by four. Then, after a longer pause, we send out first three, then three, then nine, then go back to the first group, and so on. We would have the following messages: 2,2,4 . . . 3,3,9 . . . 2,2,4 . . . 3,3,9 . . . and so on.

If then, from somewhere out in space, we got the message, 4,4,16, even once, we would have successfully established communication.

Or we might try the universal language of chemistry. There are a fixed number of types of stable atoms that should be the same all over the universe. Each different type is made up of a definite combination of two kinds of particles: protons and neutrons.

The simplest, hydrogen-1, is composed of a single proton, while the next, hydrogen-2, has a proton plus a neutron. We can therefore send out numbers representing the structure of the different atoms in order of increasing complexity. We could start with hydrogen-1 (1) and hydrogen-2 (1-1). We could then go on to helium-3 (2-1), helium-4 (2-2), lithium-6 (3-3) and lithium-7 (3-4).

Suppose then we repeat the number combination 1 . . . 1-1 . . . 2-1 . . . 2-2 . . . 3-3 . . . 3-4 . . . over and over again. Some alien intelligence receiving this series of number combinations might recognize it as representing the structure of the first few simple atoms and return signals for the next atoms in line: beryllium-9 (4-5) and boron (5-5). If they did so, we would have established communication.

Or we might try the geometric approach. We might send out a string of rapid pulses among which there was, periodically, a pulse of a special kind. There would follow a pause, then another string, and so on. Each string would have a somewhat different pattern of special pulses.

If the strings are recorded one under the other, the special pulses might combine to form a circle or some other pattern. In this way, simple geometric theorems could be transmitted; a right triangle with squares built on each side would indicate that the square of the hypotenuse is equal to the sum of the squares of the legs.

Even simple cartoons might be sent in this way, cartoons which might indicate that human beings had four limbs and stood on two of them; that they existed as two sexes, and so on. If the answer came in similar cartoons we would *really* be in communication.

Such communication would be exceedingly slow, of course, since a planet capable of answering could be anywhere in our galaxy, up to thousands of light-years away. Suppose that the intelligence we detect is 500 light-years away, a supposition which, if anything, errs on the side of optimism.

In that case, radio waves, or any other conceivable form of information-carrying signal, must take fully 500 years to travel from here to there. Another 500 will pass before the answer will travel from there to here.

Of what use would a dialogue be in which individual remarks take place at intervals of a thousand years?

In the first place, the mere fact that the dialogue exists at all would be of tremendous importance. Mankind will know itself *not* to be the only intelligence or even (very likely) the greatest intelligence in the universe, and this is bound to have a profound effect on religion and philosophy, and on our very approach to the world about us.

In the second place, neither we nor they need wait for an answer to continue talking. We can well vary our messages at will once we have established our communication. They will do the same and the end result will be a complex conversation consisting of comments intended for answer in the future, and answers for comments in the past.

Nor will the wait have to be useless. It can be extremely fruitful. If we send out simple cartoons, we can accompany each with the equivalent of a Morse code signal. A cartoon of a man would be accompanied by MAN. Men in different attitudes could be MAN WALK, MAN STAND, and whatever else ingenuity might suggest.

In 500 years we could send out a great many signals and if the intelligence we reach is superior to our own, there should be no trouble in their breaking our code. Given a certain vocabulary to begin with, they may even require no further pictures, but be able to deduce the meaning of words they don't understand from words they do.

OTHER LIFE

When the 500 years are up and they start responding, we
may well find that they have caught on quite quickly and
after only a single century, perhaps, they will have switched
to straight English. (Or straight Russian, perhaps?)

It is possible that even the simple forms of communication
with which intelligences must start may yet serve as cross-
fertilization in the realm of ideas. If we list the proton-neutron
combinations of the atoms, they may respond, eventually,
with a somewhat different listing of the atoms and in puzzling out the new listing we might, conceivably, see a regularity that now escapes us.

It is not even necessary to suppose direct and specific
information. The mere fact of interstellar communication
may help advance our technology. The effort to send out
stronger and stronger beams with greater efficiency, or to
detect weaker and weaker ones, will encourage advances
that may have application in fields far removed from that
of interstellar communication.

Then, too, the effort to concentrate as much information
into as few symbols as possible will encourage us to concentrate even more intensely on information theory. In
attempting to reach the alien minds of intelligent beings
many light-years away, we may better fit ourselves for com-
municating with dolphins here on earth. More important
still, man may even learn how to communicate more effectively with his fellow man. That consequence alone would
justify almost any conceivable effort put into an attempt to
contact aliens.

One question remains: Is it dangerous? Is it wise to draw
the attention of some supercivilization to ourselves? What
if chimpanzees somehow drew our attention to a fertile continent on which they were the highest form of life? Would
we not take over that continent, wiping out the chimpanzees
without a qualm if we felt like doing so?

Well, 500 light-years is a long distance to cross at any
level of technology, for every crossing would take an absolute minimum of 500 years of earth time. Distance alone
would probably save us.

Then, too, why ought we to be so certain that an alien
intelligence would find nothing better to do than to destroy

us? Even we ourselves, a species capable of perpetrating the Nazi horrors, have reached the point where many of us feel twinges of regret over the extinction of any form of even nonintelligent life, and would go to great lengths to prevent unnecessary interference with chimpanzees in their native habitat. Are supercivilizations to be less decent than our imperfect selves? No! I rather feel that a contact of minds across the great gaps of space could result only in good, not in evil.

Chapter 23   Anatomy of a Martian

Conditions are so different on Mars and—to our earth-centered feelings—so inferior from those on earth that scientists are confident no intelligent life exists there. If life on Mars exists at all (the probability of which is small, but not zero) it probably resembles only the simplest and most primitive terrestrial plant life (see Chapter 20).

Still, even granted that the likelihood of complex life is virtually nonexistent, we can still play games and let our fancy roam. Let us suppose that we are told flatly: "There is intelligent life on Mars, roughly man-shaped in form." What reasonable picture can we draw on the basis of what we now know of Mars—bearing always in mind that the conclusions we reach are not to be taken seriously, but only as an exercise in fantasy?

In the first place, Mars is a small world with a gravitational force only two-fifths that of earth. If the Martian is a boned creature, those bones can be considerably slenderer than ours and still support a similar mass of material (an inevitable mechanical consequence of decreased weight). Therefore, even if the torso itself were of human bulk, the legs and arms of the Martian would seem grotesquely thin to us.

Objects fall more slowly in a weak gravitational field and thus the Martians could afford to have slower reflexes. Therefore, they would seem rather slow and sleepy to us (and

This article was first published under the title of "Anatomy of a Man from Mars" in *Esquire Magazine,* September 1965.

they might be longer-lived because of their less intense fight with gravitation). Since things are less top-heavy in a low-gravity world, the Martian would probably be taller than earth people. The Martian backbone need not be so rigid as ours and might have two or three elbowlike joints, making stooping from his (possible) eight-foot height more convenient.

The Martian surface has been revealed by the Mars-probe, Mariner IV, to be heavily pockmarked with craters, but the irregularities they introduce are probably not marked to a creature on the surface. Between and within the craters, much of the surface is probably sandy desert. Yellow clouds obscuring the surface are occasionally detected and, in the 1920s, the astronomer E. M. Antoniadi interpreted these as dust storms. To travel over shifting sands, the Martian foot (like that of the earthly camel) would have to be flat and broad. That type of foot, plus the weak gravity, would keep him from sinking into the sand.

As a guess, the feet might be essentially triangular, with three toes set at 120° separation, with webbing between. (No earthly species has any such arrangement, but it is not an impossible one. Extinct flying reptiles, such as the pterodactyl, possessed wings formed out of webbing extending from a *single* line of bones.) The hands would have the same tripod development, each consisting of three long fingers, equally spaced. If the slender finger bones were numerous, the Martian finger would be the equivalent of a short tentacle. Each might end in a blunt swelling (like that of the earthly lizard called the gecko), where a rich network of nerve endings, as in human fingertips, would make it an excellent organ for touching.

The Martian day and night are about as long as our own, but Mars is half again as far from the sun as we are, and it lacks oceans and a thick atmosphere to serve as heat reservoirs. The Martian surface temperature therefore varies from an occasional 90° Fahrenheit, at the equatorial noon, down to a couple of hundred degrees below zero, by the end of the frigid night. The Martian would require an insulating coating. Such insulation might be possible with a double skin; the outer one, tough, horny, and water impervious, like that of an earthly reptile; the inner one, soft, pliable, and richly set with blood vessels, like that of an earthly man. Between the two skins would be an air space which the Martian could inflate or deflate.

At night the air space would be full and the Martian would appear balloonlike. The trapped air would serve as an insulator, protecting the warmth of the body proper. In the warm daytime, the Martian would deflate, making it easier for his body to lose heat. During deflation, the outer skin would come together in neat, vertical accordian pleats.

The Martian atmosphere, according to Mariner IV data, is extremely thin, perhaps a hundredth the density of our own and consisting almost entirely of carbon dioxide. Thus, the Martian will not breathe and will not have a nose, though he will have a strongly muscled slit—in his neck, perhaps—through which he can pump up or deflate the air space.

What oxygen he requires for building his tissue structure must be obtained from the food he eats. It will take energy to obtain that oxygen, and the energy supply for this and other purposes may come directly from the sun. We can picture each Martian equipped with a capelike extension of tissue attached, perhaps to the backbone. Ordinarily, this would be folded close to the body and so would be inconspicuous.

During the day, however, the Martian may spend some hours in sunlight (clouds are infrequent in the thin, dry Martian air) with his cape fully expanded, and resembling a pair of thin, membranous wings reaching several feet to either side. Its rich supply of blood vessels will be exposed to the ultraviolet rays of the sun, and these will be absorbed through the thin, translucent skin. The energy so gained can then be used during the night to enable the necessary chemical reactions to proceed in his body.

Although the sun is at a great distance from Mars, the Martian atmosphere is too thin to absorb much of its ultraviolet, so that the Martian will receive more of these rays than we do. His eyes will be adapted to this, and his chief pair, centered in his face, will be small and slitlike to prevent too much radiation from entering. We can guess at two eyes in front, as in the human being, since two are necessary for stereoscopic vision—a very handy thing to have for estimating distance.

It is very likely that the Martian will also be adapted to underground existence, for conditions are much more equable underground. One might expect therefore that the Martian would also have two large eyes set on either side of his head, for seeing by feeble illumination. Their function would be chiefly to detect light, not to estimate distance, so they can be set at opposite sides of the head, like those of an earthly dolphin (also an intelligent creature) and stereoscopic vision in feeble light can be sacrificed. These eyes might even be sensitive to the infrared so that Martians can see each other by the heat they radiate. These dim-vision eyes would be enormous enough to make the Martian face wider than it is long. In daytime, of course, they would be tightly closed behind tough-skinned lids and would appear as rounded bulges.

The thin atmosphere carries sound poorly, and if the Martian is to take advantage of the sense of hearing, he will have to have large, flaring, trumpetlike ears, rather like those of a jackrabbit, but capable of independent motion, of flaring open and furling shut (during sandstorms, for instance).

Exposed portions of the body, such as the arms, legs, ears, and even portions of the face which are not protected by the outer skin and the airtrap within, could be feathered for warmth in the night.

The food of the Martian would consist chiefly of simple plant life, which would be tough and hardy and which might incorporate silicon compounds in its structure so that it would be gritty indeed. The earthly horse has teeth with elaborate grinding surfaces to handle coarse, gritty grass, but the Martian would have to carry this to a further extreme. The Martian mouth, therefore, might contain siliceous plates behind a rounded opening which could expand and contract like a diaphragm of a camera. Those plates would work almost like a ball mill, grinding up the tough plants.

Water is the great need. The entire water supply on Mars is equal only to that contained in Lake Erie, according to an estimate cited by astronomer Robert S. Richardson. Consequently, the Martian would hoard the water he consumes, never eliminating it as perspiration or wastes, for instance. Wastes would appear in absolutely dry form and would be delivered perhaps in the consistency, even something of the chemical makeup, of earthly bricks.

The Martian blood would not be used to carry oxygen, and would contain no oxygen-absorbing compound, a type of substance which in earthly creatures is almost invariably strongly colored. Martian blood, therefore, would be colorless. Thus the Martian skin, adapted to ultraviolet and absorbing it as an energy source, would not have to contain pigment to ward it off. The Martian therefore would be creamy in color.

The extensible light-absorbing cape, particularly designed for ultraviolet absorption, might reflect longwave visible light as useless. This reflected light could be yellowish in color. This would cause our Martian to seem to be (when he was busily absorbing energy from solar radiation) a dazzling white creature with golden wings and occasional feathers.

So ends our speculation—in a vision of Martian forms not so far removed from the earthman's fantasies of the look of angels.

Chapter 24   On Flying Saucers

Because I frequently indulge in speculations concerning the possibility of extra-terrestrial life (see Chapters 20 to 23 inclusive) and because I am known to be a science fiction writer, I am frequently asked if I "believe" in flying saucers. The clear expectation of the individual asking the question is that *of course* I do. And by "believing" in flying saucers, the questioner usually means considering them to be space vehicles operated by nonhuman intelligences.

Well, let me make my position clear, since I don't want my writings to be used as a basis for a point of view that I consider folly.

I *don't* believe in flying saucers in the sense of considering them to be space vehicles guided by extra-terrestrials. As I explained in the previous chapters there is virtually no likelihood of intelligent life elsewhere in the solar system, and

This short item was written especially for this book.

the nearest samples of life capable of handling spaceships must occur many light-years away.

To say that intelligent life undoubtedly exists somewhere in the depths of space (as I firmly believe) is *not at all* the same thing as saying that these forms of intelligent life are visiting us in great swarms in spaceships disguised as flying saucers which are continually being sighted but which never make indisputable contact.

The energy requirements for interstellar travel are so great that it is inconceivable to me that any creatures piloting their ships across the vast depths of space would do so only in order to play games with us over a period of decades. If they want to make contact, they would *make* contact; if not, they would save their energy and go elsewhere.

Undoubtedly there are many sincere people who sight perfectly legitimate and unusual phenomena. These may not be spaceships (in fact, I am sure they are not), but there are many things other than spaceships that deserve investigation. Undoubtedly, also, scientists would react with more enthusiasm and investigate with greater energy, if past experience had not told them that the history of the flying saucer rage is full of hoaxes and frauds and error. That is not *their* fault, you know.

Therefore, without casting any aspersions on anyone, I must maintain that until an actual spaceship with its non-human crew is exhibited in the metal and flesh (lights in the sky, however mysterious, are not enough), I will continue to assume that every reported sighting is either a hoax, a mistake, or something that can be explained in a fashion that does not involve spaceships from the distant stars.

CONCERNING THE MORE OR LESS UNKNOWN

B
FUTURE LIFE

Chapter 25   The World of 1990

Predicting the future is a hopeless, thankless task, with ridicule to begin with and, all too often, scorn to end with. Still, since I have been writing science fiction for over a quarter of a century, such prediction is expected of me and it would be cowardly to try to evade it.

To do it safely, however, I must guess as little as possible, and confine myself as much as possible to conditions that will *certainly* exist in the future and then try to analyze the possible consequences. Consider, for instance, our planet's population.

There are now rather more than three billion people on earth. For the three leading nations of the world, the population figures are now roughly 700 million for China; 250 million for the U.S.S.R., and 200 million for the United States.

What will the situation be a generation from now, say in 1990, assuming that we avoid a thermonuclear war? It is virtually certain that the population will have increased by at least 60 percent. The population of the United States, for instance, may have reached the 320,000,000 mark.

Very well, then, let's get down to cases. How will everyday life here in America be lived in 1990 in the light of the population explosion? An obvious consequence is an overwhelming appreciation of the necessity of conserving the planet's resources—not out of idealism, but out of sheer self-love.

Air is inexhaustible, for instance, but to be useful it must be clean. The problem of polluted air is already serious and, by 1990, it will be an unthinkable to dump untreated smoke and exhaust into the atmosphere as it is now unthinkable to dump sewage into a city reservoir.

It is possible that this will impinge on the average human being most directly in the form of bans on smoking in the open air. It will probably be discovered that air pollution

This article appeared in *The Diners' Club Magazine*, January 1965.

(including the tobacco smoke discharged from the lungs of hundreds of millions of smokers) contributes to lung and skin cancer even among non-smokers. Smoking may therefore be restricted to "smokatoriums" where smokers can give themselves and each other lung cancer without affecting the rest of the population.

By 1990 more and more apartments will be outfitted with devices for circulating filtered air. The old-fashioned phrase "fresh air" will be replaced, perhaps, by the phrase "raw air" and this will be considered increasingly unsuitable for delicate lungs, especially in urban areas.

Again, water is inexhaustible, but fresh water is not. The squeeze for fresh water is on already. However, it seems quite likely that before the quarter of a century is up there will be practical methods for desalinizing ocean water so that, in principle, fresh water will be inexhaustible. But desalinized ocean water is bound to be considerably more expensive than natural fresh water. It will still be too expensive in 1990 for any use other than drinking or cooking and the fight against water pollution will have become strenuous indeed.

Energy sources will not yet present a serious problem in 1990. With luck, this may be no problem at all. Oil and coal will still be with us and nuclear fission plants will have become common. The great problem of disposing of atomic wastes safely will, in all likelihood, be solved. (My guess is that it will be done by mixing the wastes into glass blocks which can then be dumped into salt mines or the ocean deeps.) There will even, I suspect, be an experimental power plant or two, based on hydrogen fusion, somewhere on the planet, and considerable talk about solar power plants.

Minerals are less easy to be optimistic about. The world need will rise sharply and some mineral supplies are already critically short. Great sources, as yet untapped, are the bottoms of the continental shelves where, in some cases, nodules of metal compounds lie for the taking. Ocean dredges should be exploiting this resource by 1990.

What will impinge most directly upon the average man, however, will be the pressures on soil and living space. There are no easy solutions to the problem of crowding more and more people into the cities, but I think that by 1990 we will plainly see the direction of forthcoming change. The movement will be no longer upward into skyscrapers as has been true for the last two generations, but downward. This is not necessarily a welcome thought to those used to living in the open, but it may be inevitable and people will come to see advantages in it.

People already work and live in beehives, surrounded at all times by artificial light and conditioned air. They would scarcely know the difference if they were suddenly transported underground. Consider, too, that underground temperature changes are minor so there would be less problem of cooling in the summer or warming in the winter. If a whole city were built underground, then transportation would never be upset by rain or snow. Production would increase in efficiency, since round the clock shifts would be easy to set up in surroundings in which the difference between day and night is minimized.

Furthermore, the earth's surface will not be directly encumbered by the city. The ground above a large city may be devoted in part to park land for recreation and in part to farming or grazing. However, even in 1990, this kind of plan will still be in the reasonably distant future. Increasing numbers, though, of individual houses and factories will be built underground by then.

The population pressure will make the less desirable areas of the earth's surface seem more desirable, particularly for people who wish to get away from crowds. Those who can afford it will retreat into the isolation of the mountains, where the transportation and communication methods of 1990 will keep them in touch with humanity without subjecting them to physical crowding.

The jungle will have fewer terrors, for the large predators will be either extinct or definitely on the road to extinction by then, and the more deadly insects, worms, and microorganisms will be under better control. Increasing use of nuclear energy will begin to open the arctic coastlines and point the way to the colonization of even the empty continent of Antarctica.

Most startling of all, possibly, will be the beginning of a movement in the direction of the continental shelves. Many of the advantages of living underground can be duplicated underwater, with the added convenience that those fond of water sports need only step out of doors. The chance of catching dinner in your front yard (once possible for early Americans) will be possible again underwater. Perhaps in 1990 a large hotel will be in the process of construction underwater—off Miami, Florida, I suspect.

Population pressure will not yet have driven men off the planet altogether. There ought to be a going colony on the moon, composed of changing shifts of highly trained and specialized personnel, and there will be plans for landing a man on Mars. The average man, nevertheless, will be as far removed from the chance of a trip into space in 1990 as he is now. But the space age will be far advanced just the same (see Chapter 30).

The major human conglomeration of 1990, despite the starts made toward the underground and the continental shelves—and even the moon—will still be in the same large cities we have today; cities which will, by then, be enormously larger. The northeastern seaboard of the United States will be, in effect, a single large city of about 40 million.

To keep the millions of city dwellers comfortable, there will have to be considerable refinements in transportation and communications. Garages will proliferate and become lavish, jutting both aboveground and below. Their effectiveness will be accentuated by the growing use of two-seater runabouts for intracity use. (I suspect that excise taxes will rise steeply with over-all car measurements in an effort to encourage the use of extra-small compacts.)

Personal vehicles will be separated from commercial vehicles as much as possible. The elevated street will become an increasingly common sight in the congested centers of the huge cities and will be used by the small cars, while buses and trucks will be confined to what is now street level.

Helicopter deliveries of nonbulk items will achieve limited popularity. The newer buildings of 1990 will be topped by small heliports, perhaps as much for show and prestige as for use. There will also be an increasing tendency to make use of tubes and compressed air for mail service. The post office will be extensively automated. I suspect that large office buildings, at least, will have their mail delivered by puffs of air, with it then being rerouted to individual suites with a minimum of human touch.

Subways, too, will become increasingly automated and there will be a strong trend, by 1990, toward continuous chain subways—a long series of coaches traveling the length of a line and back circling wide at either end. This will still be confined to small shuttle lines, but engineers will be presenting designs for city-wide affairs of this sort, with highly controversial solutions for such matters as getting on and off a continuously moving chain and for methods of interconnecting the separate chains.

Between the cities the steady decline of the railroad will have produced trucks and buses of unprecedented size and sorts. More and more they will come in tandem and highways will have to take these monster vehicles into account. They will have their special lanes and their special entrances and exits.

For other than commercial use, 1990 may find the intercity highway passing its peak. The use of personal helicopters will increase but even more so, perhaps, will the ground effect vehicle come into its own. The latter, running on jets of compressed air rather than on wheels, will not require paved highways, but will be able to move along dirt roads with equal ease or, for that matter, across open country (if rendered not too uneven by either natural or man-made obstacles) and bodies of water.

The ground effect vehicle will undoubtedly require drastic changes in traffic regulations. One of the growing irritations of 1990 will be the disregard of drivers of such vehicles (particularly teen-agers) for private property rights. I imagine there will be a tendency for irascible landowners to raise deliberate obstacles to discourage this, and, if a youngster is killed because of such an obstacle, a pretty legal hassle will ensue.

Perhaps the most forceful effect of the population explosion will be in connection with food. The United States will not be experiencing the famines that will be all too common in much of the world, but we will have to grow more food conscious and less food particular. There will be an increasing tendency to grow less specialized in our diet, as well as a tendency to drift away from meat and toward fish and grain.

Items not now considered palatable will be entering the dietary, though only on an experimental basis (for it takes the threat of actual starvation to make a population give up its food prejudices—and sometimes not even that). Seaweed is one example of a food which may reach the restaurants. There will also be increasing experimentation with cultured algae and yeast. The supermarkets will be stocked with such items, artificially flavored to resemble meat, liver or cheese, I daresay. In 1990, these artificial flavors will still undoubtedly leave something to be desired.

One major item other than the population explosion that seems certain to come to pass is the continued push toward an extreme of mechanization and automation. This will be particularly true in the United States, which, of course, will continue to live its gadget-centered existence.

This will affect the housewife, from shopping through final consumption. The supermarket of 1990 will have its items coded. The shopper will mark off the code numbers of desired items on appropriate cards, using shielded display counters as a guide. Her order, properly packed, itemized, and charged, will be waiting for her within minutes.

Most food items will be prepared for cooking in kitchen units that will do the job with a minimum of human interference. The kitchen may come to resemble the cockpit of a jet bomber. In fact, there will be apartment houses in 1990 that may very well offer a community kitchen for the use of their tenants (as they now offer community laundries) since that will eliminate the wastefulness of elaborate units for each apartment. (Even so, kitchenettes for preparation of breakfasts and snacks would remain in each apartment.) The trend would definitely be for "restaurant eating" even at home.

The "servant problem" will continue insoluble in the United States and the substitution of the household robot will not alleviate the situation. What will alleviate it will be the increasing tendency to reduce the chores requiring servants (or housewifely muscles). The increasing use of filtered

air will make the dust problem smaller. Washing by ultra-
sonic vibrations in addition to (or in place of) soap will
make that task much quicker and easier.

Automation will bring about a change in work outside
the home. More and more, the sheer use of muscles or the
routine use of brain will vanish. There always will be creative occupations, of course, as well as a need for executives,
administrators, and for all people who must deal with other
people. There will also be a tremendous increase in the number of people who must, in one way or another, deal with
computers and their offshoots.

For that reason, education will feature mathematics and
science more and more. Such items as binary arithmetic and
computer languages will be taught from the earliest grades.
Personalized education and detailed teacher-student contact
will tend to restrict itself to two classes of children—the retarded and the very bright.

The greatest single problem introduced by automation will
be surplus time. The large majority will be working only 30
hours a week at most, and will therefore be more subject than
ever to the dangerous disease of boredom. There will have
to be a great emphasis on recreation and entertainment,
and never in the history of man will so great an importance
be attached to the general profession of "people-amusing."

The television set will be, more than ever, the center of
the home, and the telephone itself will become almost an item
of entertainment. The 1990 phone will be routinely equipped
with a television attachment so that one can see as well as
hear the person at the other end. The housewife can then
enjoy herself twice as much, provided she is in a condition
to be seen and is willing to use the vision attachment. (A
whole new dimension of strain among friends will arise when
one is highly polished and lacquered and wishes to see and
be seen, while the other is hung over and wishes to do
neither.)

Such a telephone may also revolutionize library work. By
1990 the large libraries will have all but current and popular
books on microfilm. All schools and many homes will have
microfilm viewers. The large libraries may well be organized
to allow telephonic viewing of their microfilms. It will then

be possible to check references and obtain information without leaving home or office.

The businessman may view documents and receive reports by "visiphone." There may even be occasions when conferences by split screen telephone can be arranged. The money expended can be saved on the travel that will not be necessary (except for the large percentage of cases where the travel is an excuse for a junket at the general expense of society).

Sports also will be stressed in the world of 1990 as a good and harmless time consumer. I suspect that the great sports novelty will be flying. Small motors, mounted on the back, will lift a man clear of the ground. By 1990 this should still not be cheap enough and common enough for transportation, but should be adequate for thrills and sport. (Will some child, alive today, be the man who will organize the first game of "air-polo" using a helium-inflated sphere of thin but tough plastic as the ball?)

The changes that will be taking place between now and 1990 will convince people that the trend cannot be allowed to continue blindly, but must be deliberately channeled. There are many today who are convinced that effective birth control of some sort is essential if civilization is to be saved. They are in the minority now—but they won't be in 1990.

By 1990, in fact, governmentally organized measures for birth control will be taken for granted over almost all the world. The advance in birth control effectiveness will not be in time to prevent the 60 percent increase in world population by 1990, but it will have reached the point where the percentage of individuals under 21 will be markedly smaller than it is today.

This should bring about a change in the social attitude toward children and family, though this change might not be uniform everywhere. In some areas and among some segments of society, the relatively small number of children may increase the value of those who do exist, thus making that society more child-centered. In other areas and segments, the recognition of the population explosion as the prime danger to man may make children unpopular and parenthood seem vaguely antisocial. Family bonds may tend to dissolve and marriage may lose ground to other less formal types of personal union.

And if 1990 further sees the beginning of a population equilibrium or even a population rollback, the writer of that day forecasting the world of 2090 may have grounds for considerable optimism.

Chapter 26   The World's Fair of 2014

The New York World's Fair of 1964-65 was dedicated to "Peace Through Understanding." Its glimpses of the world of tomorrow rule out thermonuclear warfare. And why not? If a thermonuclear war takes place, the future will not be worth discussing. So let the missiles slumber eternally on their pads and let us observe what may come in the non-atomized world of the future.

What is to come, through the Fair's eyes at least, is wonderful. The direction in which man is traveling was viewed with buoyant hope, nowhere more so than at the General Electric pavilion. There the audience was whirled through four scenes, each populated by cheerful, lifelike dummies (including a dog that steals the show).

The scenes, set in or about 1900, 1920, 1940, and 1960, show the advances of electrical applicances and the changes they are bringing to living. And if they had gone on to include 1980, 2000, and so on, what would they have showed? I don't know, of course, but I wonder—

If we consider some of the changes mentioned in Chapter 25, and some not mentioned there, what will the World's Fair of 2014-15 turn out to be like?

One high-probability development is that men will continue to withdraw from nature in order to create an environment that will suit them better. By 2014, electroluminescent panels will be in common use. Ceilings and walls will glow softly, and in a variety of colors that will change at the touch of a push button.

A longer version of this article appeared under the title of "Visit to the World's Fair of 2014" in *The New York Times Magazine*, August 16, 1964.

Windows need be no more than an archaic touch, and even when present would be polarized to block out the harsh sunlight. The degree of opacity of the glass may even be made to alter automatically in accordance with the intensity of the light falling upon it.

This will still be a luxury development and most ordinary mortals will not have such devices in their own homes. The Fair of 2014, however, may well prove a symphony in electroluminescence, with scarcely a real window in any structure.

There was an underground house at the 1964 Fair which seems to me to be a sign of the future. If its windows are not polarized, they can nevertheless alter the "scenery" by changes in lighting. There are certain advantages to underground living (see Chapter 25) and at the Fair of 2014, General Motors' "Futurama" may well display vistas of underground cities complete with light-forced vegetable gardens.

Gadgets will continue to relieve mankind, increasingly, of tedious jobs, and the final third of the 20th century should see the arrival of the household robot. Robots will be neither common nor very good in 2014, but they will be in existence.

The IBM building at the 2014 World's Fair may have, as one of its prime exhibits, a robot housemaid—large, clumsy, slow-moving, but capable of general picking up, arranging, cleaning, and manipulation of various appliances. It will undoubtedly amuse the fairgoers to scatter debris over the floor in order to see the robot lumberingly remove it and classify it into "throw away" and "set aside." (Robots for gardening work will also have made their appearance.)

General Electric at the 2014 World's Fair will be showing 3-D movies of its "Robot of the Future," neat and streamlined, its cleaning appliances built in, and performing all tasks briskly. (There will be a three-hour wait in line to see the film, however, for some things never change.)

By the opening of the 21st century, man's energy needs will be largely met by nuclear sources even in small ways. The applicances of 2014 will have no electric cords, for instance, but will be powered by long-lived batteries running on radioisotopes. The isotopes will not be expensive for they will be by-products of the fission-power plants which, by 2014, will be supplying well over half the power needs of humanity.

An experimental fusion-power plant or two will already exist in 2014, and it is in this direction that the 2014 Fair will point its eyes. Even the World's Fair of 1964 was able to demonstrate a small but genuine fusion explosion—but the 2014 Fair will show models of advanced fusion plants and will have devices that will allow the production of enough electric power to keep displays in operation. ("This Electricity Produced Through Fusion.")

By 2014, large solar-power stations will be in operation in a number of desert and semi-desert areas—Arizona, the Negev, Kazakhstan—where sunlight is reliable and steady. In the cloudy and smoggy areas of the crowded cities, however, solar power will be less practical and the attempt to shift the collection of such power to space will be well-advanced. An exhibit at the 2014 Fair will show models of power stations in space, collecting sunlight by means of huge parabolic focusing devices and radiating the energy thus collected down to earth.

The world of 50 years hence will have shrunk further. At the 1964 Fair, the G.M. exhibit depicted, among other things, "road-building factories" in the tropics, and visitors to the Fair in the present could travel there on an "aquafoil," which lifts itself on four stilts and skims over the water with a minimum of friction. The mechanics of transportation, one can see, is advancing rapidly and will continue to advance.

By 2014, it seems to me, much effort will be put into the designing of vehicles with "robot-brains"—vehicles that can be set for particular destinations and that will then proceed there without interference by the slow reflexes of a human driver. I suspect that one of the major attractions of the 2014 Fair will be rides on small roboticized cars which will maneuver in crowds at the two-foot level (held up by jets of compressed air, neatly and automatically avoiding each other.

For short-range travel, moving sidewalks (with benches on either side, standing room in the center) will be making their appearance in downtown sections of cities and certainly the sidewalks of the World's\Fair of 2014 will all be mechanized.

Communications will also be advanced and synchronous satellites will have made it possible to call anyone anywhere on earth with a minimum of trouble. That would not be worth attention at the Fair of 2014 so routine would it be. But the moon?

By 2014 there will certainly be a permanently manned station on the moon, and any number of simultaneous conversations between earth and moon can be handled by modulated laser beams (see Chapter 11), which are easy to manipulate in space.

If the lunar colony is in a position to co-operate, it may be possible to offer Fairgoers a chance now and then at a real conversation with a man on the moon.

Such conversations would be a trifle uncomfortable, by the way, since 2.5 seconds must elapse between statement and answer. (It takes light and radio waves that long to make the round trip.) Similar conversations with Mars would experience a 3.5 minute delay even when Mars was at its closest. However, there will probably be no chance at Earth-to-Mars conversations at the 2014 Fair. By then only unmanned ships will have landed on Mars, though a manned expedition will be in the works and the NASA display will show an elaborate Martian colony.

As for television, wall screens will have replaced the ordinary set by 2014; and transparent cubes will be making their appearance. In the videocubes, with the help of holography (see Chapter 11), three-dimensional viewing will be possible. In fact, one popular exhibit at the 2014 World's Fair will be such a 3-D TV, built life-size, in which ballet performances will be seen. The cube will slowly revolve for viewing from all angles.

One can go on indefinitely in this happy extrapolation, but all is not rosy.

As I stood in line waiting to get into the General Electric exhibit at the 1964 Fair, I found myself staring at Equitable Life's grim sign blinking out the population of the United States with the number (which was then something over

191,000,000) increasing by 1 every 11 seconds. During the time which I spent inside the G.E. pavilion, the American population had increased by nearly 300 and the world's population by 6,000.

In 2014, there is every likelihood that the world population will be 6,500,000,000 and that the population of the United States will be at least 350,000,000. Ordinary agriculture will keep up to food requirements with great difficulty, if it manages to do so at all, and there will be "farms" turning to the more efficient micro-organisms.

The 2014 Fair will show evidence of this and it will feature a Yeast Bar at which "mock-turkey" and "pseudo-steak" will be served. It won't be bad at all (if you can dig up those premium prices), but there will be considerable psychological resistance to such an innovation.

It will be overwhelmingly obvious by 2014 (see Chapter 27) that the population explosion cannot be allowed to continue unchecked for much longer. Already, birth control measures will be in use everywhere and there will be a strong effort toward making them still more popular.

One of the more serious exhibits at the 2014 Fair, therefore, will be a series of lectures, movies, and documentary material at the World Population Control Center (adults only; special showings for teen-agers).

And on the success of that will rest the chances of having any Fair at all in 2064—or perhaps, even, any civilized world.

Chapter 27   Fecundity Limited

In the previous two chapters I have hinted at the disasters that await mankind if the present rate of increase in population is allowed to continue indefinitely. There are many, however, who take a casual view of the matter and assume that "science" will always find a way; that no matter how large our numbers become, scientific advance will discover a way to feed, house, and amuse us.

Is that so?

Let us ask then: How far *can* we increase our numbers on earth and how long will it take us to reach our limit?

Let's give mankind every possible break in this matter so that there is no question of being anything but supremely optimistic. Suppose energy to be no problem; hydrogen fusion and solar power to give us all we need. Suppose we have worked out artificial photosynthesis and can form all the food we want as fast as we want out of water and air, as plants now do. Suppose we solve all the organizational problems of dealing with a tremendously crowded planet (from coping with waste disposal to the handling of racial tensions). Suppose, even, that we can wipe out all competing life to make the maximum room for ourselves.

If we suppose all that, what can limit man's population increase? Well, one thing cannot be avoided if we are restricted to our own planet. Sooner or later, we will run out of at least one of the chemical constituents of the human body—we get to the point where there just isn't enough left anywhere on earth to make another human being.

Actually, the element which is in the most critical supply, and which is the one that will probably be used up first if mankind increases without limit, is phosphorus. However, let us give mankind a further break by considering carbon, a less critical component of life from the standpoint of sheer mass-availability, and see what conclusions we can come to.

To begin with, not all the carbon on earth is in a form that is readily available to life forms. Let us begin then with merely the "available carbon."

Ninety percent of the available carbon occurs in the ocean as bicarbonate ion. A small amount is in the air in the form of carbon dioxide, and the rest is contained in living creatures. You can add to this the ordinarily unavailable carbon content of earth's oil and coal, since these are being rapidly burned and converted to carbon dioxide, which enters the air or dissolves in the sea and becomes available to life.

The total amount of carbon present on earth in these forms comes to about 51,000,000,000,000,000,000 grams (which is equivalent to about fifty-six trillion tons).

This is truly a sizable quantity but wait, some 90 percent of that carbon must be reserved for man's food supply (assuming that he isn't reduced to cannibalism). After all, man must eat, and he must eat carbon-containing food, whether it is grown in the soil or in chemical tanks, whether it is meat, wheat, yeast, or a mixture of nutritious chemical compounds. And an over-all organic food supply ten times the mass of humanity is necessary to allow a safe margin, as well as to allow for the production of nonedible organic by-products such as textiles, plastics, and so on. That still leaves us with something over five trillion tons of carbon that can actually be incorporated into human beings.

Now let us suppose the average human being on earth (including children) weighs 100 pounds. Each one would contain an amount of carbon coming to 18 percent of his total weight. This would be 18 pounds or some 8,100 grams. The number of human beings required, then, to exhaust ten percent of the available carbon on earth would be 630,000-000,000,000.

This number, six hundred and thirty thousand billion, certainly dwarfs our present population of a mere three billion and makes it seem that we have ample time in which to expand; that the problem of pushing near the maximum potential is far, far in the future. But is it?

Earth is currently doubling its population every half-century but let us be conservative and say it is doubling its population every 80 years. If this rate of doubling continues, then in about 1500 years, that is, by 3500 A.D., we would have reached maximum. The living matter on earth would then consist only of human beings, and their necessary supply of food and organic by-products.

If earth's population were to spread itself evenly over earth's land area, each person would, in 3500 A.D., have about 2½ square feet to stand on, and this includes Greenland, Antarctica, the Amazon Valley, the Sahara Desert. That's what I call crowding.

I think you will agree that no reasonable extension of scientific ability will make such a condition tolerable; or even a condition approaching it. Therefore, if the population explosion continues unchecked, an intolerable crisis will be upon us in *much less time than 1500 years, no matter what science does.*

But just for fun let us suppose that through some unimaginable scientific advance, even this population can be taken care of. What next?

Well, as I said earlier, there is far more carbon on earth than is ordinarily available. There is the carbon tied up in limestone and other materials making up the crust of the earth. This carbon is not generally available to living creatures until slow geologic processes move it into air or sea. But let us be optimistic. Let us assume that mankind can burrow as far as necessary into the crust and make all of the carbon available.

The quantity of the carbon in the earth's crust is nearly 500 times the amount available in air and sea, so mankind might be envisaged as multiplying to 500 times the 3500 A.D. population.

This would bring the total population of the earth to 300,000,000,000,000,000 or three hundred million billion. If this mighty number were spread over the earth (and this time, we can suppose the oceans to be covered over with planks from end to end so that people could stand on them, too) each individual would have about one-eighth of a square foot to stand on. They would have to be stacked like cordwood.

And how long would it take for humanity to incorporate all the carbon on earth, available and unavailable, into their bodies and their food? Only an additional seven centuries after the 3500 A.D. mark. In 4200 A.D. there would be an absolute end.

But, then, why restrict ourselves to our own puny planet of earth? The Space Age has come upon us. Science is making tremendous strides. Out there the vast illimitableness of space beckons. Surely there is room out there for any number of human beings, and we don't have to worry about population expansion.

Don't we?

There are 135,000,000,000 stars in our galaxy and perhaps 100,000,000,000 galaxies in the known universe. Let's suppose that every known star in the universe is surrounded by ten planets, each capable of supporting as much life as earth can.

Let's suppose moreover that there is no problem whatever

in moving earthmen to any planet in the universe at a moment's notice. Just snap your fingers and it is done.

Now, then, in what year will all the universe be filled, to the same extent as earth of 4200 A.D.? In what year, will earthmen be stacked like cordwood over the entire surface of every one of a couple of trillion trillion planets?

Why, roughly speaking, by 11,000 A.D.

In short at the present rate of increase in population, Homo sapiens can fill the universe far beyond any question of tolerability in a mere nine thousand years.

There is no room, you see, and science can do nothing. The rate of population increase *must* decrease, and this can be done in one of two ways—either by increasing the death rate or decreasing the birth rate.

Take your pick.

Chapter 28 *The Price of Life*

Uninhibited increase in population is not the only danger that faces humanity. A more subtle one is the drive for extended lifetimes and even immortality. What if the population is stabilized in numbers but the individual lives forever?

There are organizations in being now that aim to organize the deep freezing of freshly-dead or about-to-be-dead bodies. The idea is to revive the frozen bodies when science has learned how to cure the disease that has killed them, reconstruct their broken bodies, reverse old age, restore life. We will then each one of us be a Lazarus.

Why not? What do we have to lose? If science never learns how to restore us to life, youth, and health, we are no deader than we would have been anyway, and at least we died with hope. If science learns, then we are essentially immortal.

Who can complain about a game in which the possible gain is infinite and the possible loss is nothing? The funny part is, I can. For it is the gain that is nothing and the loss that is infinite.

In saying this, I am not thinking of the individual, although even in his case, immortality is not what it might seem to be. After all, who has ever pictured a really attractive heaven?

It might be nice to put on a white robe and a halo and fly over golden streets all day, and sing hosannas and halleujahs in perfect chorus, and smile purely at the young lady angels—but *forever?* I might be able to stand it as a curiosity for a couple of days, especially the flying part, but after that I'd start drumming my fingers.

Of course, a heaven need not be puritan. What if we all ended up in Valhalla, hacking away at the giant boar and swilling mead and making love to Valkyrie after Valkyrie (or to ever-virginal houris in the Moslem heaven)? That might last a little longer but surely before much time had passed, the cry of "Pork *again?*" will resound through the gigantic hall, and the Valkyrie (or houri) of the day will be looked upon with a certain loathing.

There is a strange alchemy about forever. It can take all that is finest and best and change it into boredom. Nothing can escape. Weariness is all.

Of course, if we remain on the individual level, this problem can perhaps be solved. After all, we don't have to settle for *absolute* immortality. No one need be *forced* to stay alive.

If one wants to leave the world of the living in a society of potential immortals, one can. In such a society, the very cap and climax of life may be the civilized death. It might even be that special centers would be established where one can hold the equivalent of a convivial wake before death; one last celebration, one last clasp to the breast of the loved ones who have not yet gone before, one last shake of the hand of all the trusty friends.

Then, to the strain of soft music, and to a last burst of waving hands and finger-kissing, the compartment closes about you, the nerve gas sifts in and you are gone.

In other words, immortality means not "forever" but "as long as you want."—And how long is that?

Naturally, it varies from person to person. Somerset Maugham, who died in recent years at 91, longed for death, but he was old and sick and blind. In an immortal society, we would naturally expect to retain the vigor and strength of youth for as long as we care to live. How long will a strong young man of sensitivity and intelligence have to wait before he wishes for death to place a term to weariness?

If he is fortunate enough or able enough to work out a life in which he faces a truly challenging problem—if he directs the affairs of humanity or guides the assault of knowledge upon the unknown or distills beauty out of the universe—he is not likely to be bored quickly and may last a long time before the final wave of the hand.

Shall I guess? Five hundred years on the average?

The statesmen of the world, the scientists, the artists, the scholars will be vigorous multi-centenarians on the whole and there, exactly there, is the real danger.

An individual's brain is of prime importance to humanity only until he is thirty-five. If by then, he has not shown clear evidence of great talent, he is not likely ever to show it. If by then he *has* shown such evidence, he will probably spend the rest of his life mining the great concepts of his youth. If he were to die at thirty-five, other lesser men could mine those same concepts without much greater difficulty.

Isaac Newton was 25 when the main line of his later accomplishments was worked out in his mind. Albert Einstein was 26 when he first worked out the theory of relativity. Charles Darwin was 22 when he set out on the voyage of the *Beagle* and made those observations from which he later extracted the theory of evolution by natural selection. And so on, and so on, and so on.

This is not to say that great work is never done by oldsters (Winston Churchill's greatest accomplishment came at 65) or that there isn't an occasional "late-bloomer" (Joseph Conrad began to write when he was 37). Nevertheless, almost all the great seminal advances in human history, the great breakthroughs in new directions, have been made by young men.

And that's only natural. The human mind hardens quickly. This has nothing to do with physical deterioration of the brain or its limited capacity, and the problem won't vanish if we assume a society of immortals with brains that remain physically young. Once a brain develops a way of thought, that way wears a quick rut among the convolutions, so to speak, and it is only with the greatest effort that the line of thought can force its way out of that rut.

The great physicist, Max Planck, once said that the only way to make a startling new theory acceptable to science was to advance it, then, if it was proven useful and valid, to wait for all the old scientists to die.

It is only the young, essentially blank, mind, not yet tracked over by the muddy feet of already accomplished thought, that can see the truly revolutionary solutions. And then, in the course of a decade or so, the young revolutionary becomes the new-orthodox. How often this has happened in science, art, scholarship, and politics.

Well, then, are we to have a world in which these key fields are to be dominated by multi-centenarians in no hurry to let go?

Shall we avoid the individual death of an aging body, and the mass death of nuclear fire, only in order to sink into the slow, tired death of rust?

Death is the price we pay for meaningful life. Death makes way. Death forces the tired and old to give ground to the bright and new. Death wipes clean and prepares the ground for new advance.

But can the individual reasonably be content with individual death for the sake of abstract humanity? Why not? We expect a man to die for the sake of his family or his country. Why not for mankind? No individual lives a life he himself has built alone. All men live lives which are in their every detail the conglomerate accomplishment of the other men that live now and have lived before. The life which has been made possible to the individual by the species, he surely owes to the species.

Of course, one might draw upon the fantasy of an omnipotent science and say that, instead of suicide, a person who was tired of life might have his brain rinsed of its accumulated trackings. He might once more face the universe with a fresh mind and begin all over, like the houri of the Moslem heaven with her constantly renewed virginity.

But if you start fresh, have you not died? If a past life is not remembered, the individual represented by the past life is dead.

Then let's not go too far. We can make the rinsing of the

brain a partial one. We can leave the basic personal memories that will allow a continuity to the personality. We can perhaps leave the basic schooling that will save the necessity for a new education from scratch. We will just skim off the accumulated crud.

Unfortunately, the basic schooling already points the way; the existence of a given personality already indicates the tendency. The new individual, however freshened, is no advance on the old one, and will repeat himself in essentials forever.

Even a complete blank and the acceptance of a mental death while clinging to merely physical immortality is not sufficient. There is a basic difference between an old individual with his brain refurbished or renewed, and a completely new individual. The renewed individual is the product of *one* old individual, but the completely new individual has *two* parents.

Each child is born with half his genes from one parent and half from the other. His basic chemistry is different from either parent and (barring the case of identical multiple births) different from every other person who has ever lived. The brain of the newborn child is not merely a clear brain; it is a clear, *different* brain.

We die alone, but we are born of a pair. Sex is not only fun, but it is the method worked out through billions of years as the most effective means of maintaining the flexibility of life in the face of ever-changing environment. What we need are new and different individuals, not merely the old ones, dry-cleaned and pressed.

Yet even if we grant that immortality for the individual is death by decay and boredom for the species, may we not argue fatalistically that species die anyway so why sacrifice personal immortality for what is mortal whatever we do? To be sure, thousands of species have died despite all that sex and individual death could do.

And yet if a species dies through the halt of evolution by way of individual immortality, that is death absolute! If, on the other hand, sex and individual death are allowed to continue the evolutionary process, it is quite possible that Homo sapiens will die only after having given rise to something different and (it is to be hoped) better than himself.

239

Surely if the species must die, let it die while leaving behind a greater species that can take up more effectively the eternal struggle with darkness and stride to the kind of victories we can't even imagine today. Properly viewed, such a species death is no death at all, but another step toward the only worthwhile immortality—that of life and intelligence in the abstract.

Chapter 29   The Moon and the Future

I suppose mankind can't help taking the moon for granted. It's always been up there, playing a very soft second fiddle to the glorious sun. Its changes in phase, from new to full and back to new again, defined the original "month" and helped men devise the first calendars.

Its most marked physical effect on the earth is its ability to lift the waters of the ocean toward itself. This produces the tides, which for many centuries men seemed to blame on everything *but* the moon.

When the telescope was invented, the first heavenly object on which it was turned was the moon. It became more than a shiny object; it became a world with mountains, craters, and large, level regions that were called "seas."

But additional telescopic studies soon made it plain that the level areas were not seas and that, indeed, there was no water to speak of on the moon. No air, either.

The moon, astronomers came to believe, was a dead world; an unchanging world. It was without air and therefore without sound or weather. It was without water and therefore without life. As it was, so it had always been and so it would always be. . . . At least, that was the view put forward in the astronomy textbooks.

And now that the space age is upon us and we look forward to leaving the earth, what must be the first target for our initial rickety steps outward? Why, the moon again.

Shall we be disappointed? Are we to be bitter over the fact that billions must be spent, lives must be risked, incredi-

This article first appeared under the title "What Can We Expect of the Moon?" in *The American Legion Magazine*, March 1965.

ble effort must be exerted—and for what? To land on a barren rock, a desert, the bare, bleached corpse of a world.

And yet we should not be disappointed at all. On the contrary, we should thank the fate that seems to have designed the solar system for the express purpose of making astronauts happy.

Consider . . .

If the moon is left out of account, the nearest bodies to the earth would be the two planets, Venus and Mars. The former is never closer to us than 25 million miles, and the latter is never closer than 35 million miles.

To try, first time, to place men on worlds so distant would be such a formidable undertaking that mankind might never be able to nerve itself to try.

Fortunately, a body—the moon—has been placed much closer to us. For the moon is, on the average, only 237,000 miles away. It is a little less than 1/100 the distance of Venus at its closest, a little less than 1/140 the distance of Mars at its closest. Its distance represents less than ten times the trip around the earth at the equator. What's more, Venus and Mars reach their closest points only at intervals and remain there only briefly, while the moon is at its distance all the time.

Astronomically speaking the moon is next door, ideally placed for even the most fumbling and primitive of space shots. Thus, it is less than a decade since mankind first placed an object into orbit about the earth and already the moon has been passed, circled, photographed from far and near, and subjected to soft landings of instrumented packages.

Reaching the moon is exactly the exercise we need to develop our space muscles, to learn the proper techniques of how to live in space and on alien worlds. With the experience thus gained we will be able to learn how to reach the planets with far less difficulty than would have been our lot if we had tried to reach those same planets at one great bound.

There is the first important reason for reaching the moon. It is probably the only way in which we can learn to take further steps and enter the space age in full force.

But though we might recognize the great good value of

having the moon where it is, ought we to be amazed about it? After all, the moon does exist and it is there. Why not accept it?

The answer to that is that in studying the rest of the solar system we cannot help but come to the conclusion that the moon by rights ought not to be there. The fact that it is, is one of those strokes of luck almost too good to accept.

There are 31 known satellites in the solar system and of these fully 28 are in the possession of but four of the planets: Jupiter, Saturn, Uranus, and Neptune. These are giant planets, each of them much larger than the earth. Their gravitational fields are immense and you would expect them to hold satellites in their grip. Jupiter, the largest planet, possesses twelve known satellites. Saturn, second largest, possesses ten.

The small planets, such as earth, with weak gravitational fields, might well lack satellites. Pluto has no known satellites; neither have Mercury and Venus. (Venus is a particularly interesting case, for it is just about the size of earth, yet has no satellites. If mankind had evolved on Venus instead of on the earth, space travel might remain completely impractical.) Yet earth, quite surprisingly, does have a satellite —the moon.

But wait a bit. I haven't mentioned Mars. Mars, although only 1/10 as massive as the earth, has *two* satellites. What about that?

Well, it's not just having satellites. It's primarily the *size* of those satellites.

For instance, let's take a look at Jupiter's twelve satellites. Seven of them are tiny things, a couple of dozen miles in diameter apiece. They are probably small chunks captured by giant Jupiter out of the material of the asteroid belt that lies between itself and Mars. An eighth satellite is only 100 miles through. The remaining four, however, are large worlds, with diameters of from nearly 2,000 miles to over 3,000.

All of Jupiter's satellites put together, however, are less than 1/500 as massive as Jupiter itself. Similarly, Mars has two satellites but both are tiny things, about a dozen miles in diameter. Together they make up only about 1/500,000,-000 the mass of Mars.

In general, then, when a planet does have satellites, those satellites are *much* smaller than the planet itself. Therefore even if the earth has a satellite, there would be every reason to suspect (if we didn't know better) that at best it would be a tiny world, perhaps 30 miles in diameter.

But that is not so. Earth not only has a satellite, but it is a giant satellite, 2,160 miles in diameter. There are only seven such giant satellites in all the solar system. Monstrous Jupiter has four of them, and Saturn and Neptune have one each. Giant Uranus has none. How is it, then, that tiny earth has one? Amazing!

The moon is 1/81 as massive as the earth. No other satellite anywhere is nearly as large compared to the planet it circles as the moon is compared to the earth. Indeed, the moon and the earth form a "double planet" system that is unique in the solar system.

There's the incredible luck we have. Not only does earth possess a moon to serve as our first steppingstone into space, but it is a giant-size steppingstone that is infinitely more interesting and useful than a small object the size of a Martian moon would be.

The surface area of the moon is 14,600,000 square miles, which is about the area of Africa and Europe put together. This is a lot of room to explore.

To be sure, that surface has been photographed in front and behind, from far and near. Robot devices could easily be landed on the surface to do more than photograph it—to test and analyze it physically and chemically. You might wonder, then, why anything more is needed. Why go to the danger and expense of sending men?

Leave out of account that men will insist on going; that curiosity and the drive to brave the unknown are not to be beaten down, and there is still the fact that no instrument yet devised can match the wonderful complexity of the human brain.

We don't know what surprises may be in store for us in those 14,600,000 square miles. We don't know what some odd corner hidden in the shadow of a crater wall may reveal. Only the agile human brain can be counted upon to meet all exploration surprises properly.

Then, too, not all the aerial photographs can thoroughly reveal all the corners of the moon's vast surface. Even after we have landed on the moon, it will take decades to explore and map it completely and those will be exciting decades indeed for the brave men who will tramp the moon.

Is such exploration practical, though? Won't we be throwing away the valuable lives of our young astronauts?

Lunar exploration *is* practical. Dangerous, certainly, but in some ways not as dangerous as the exploration of the earth itself. The lunar explorers will not face hostile tribes, or dangerous animals, or deadly bacteria. There will be just the inanimate environment which, however risky, offers dangers that can be pre-calculated.

First of all, the moon is airless and waterless, but then so is outer space generally. The lunar explorers will bring water and air with them, as well as food and other necessities of life. They will be out in the open in a spacesuit which will possess an air supply, a warming unit, and other devices that will serve to make the small portion of the universe immediately next to the body safe and comfortable.

A greater danger is the sun. It is as large and as bright shining down upon the moon as upon the earth. On the moon, however, there is no air to absorb the dangerous short-wave radiation. Sunshine on the moon is therefore much richer in ultraviolet radiation and in x-rays than sunshine on earth. Fortunately, the explorer is not unprotected against radiation. Even the transparent material of his headpiece would be of a composition that is opaque to the milder types of energetic radiation. Cosmic rays are a greater problem and these might limit the amount of time an explorer could spend in the open at any one clip.

The sun's heat is an ever-pressing danger. The moon is much hotter, in spots, than the earth is, because our satellite rotates on its axis so slowly. It rotates once in 29½ days, which means that a particular spot on the moon will experience a two-week-long daylight period followed by a two-week-long nighttime period. (We see the changing pattern of light and dark in the form of the phases of the moon, which go through a complete cycle in 29 ½ days.)

During the two-week period of daytime, points on the moon's equator (which receive the most concentrated dose of sunlight) reach a temperature a bit higher than the boiling point of water. It is better for an explorer not to be at those points on the moon's surface.

Fortunately, it is easy to stay out of direct sunlight on the moon. Because of the moon's slow rotation, sunrise is very slow and the explorer is never likely to be surprised by a burst of sunlight as the night suddenly ends. At the moon's equator, the terminator (that is, the line separating day and night) moves westward at a rate of 9½ miles an hour. This motion is even slower at points far removed from the equator. At 60° North or South Latitude on the moon, the motion is less than five miles an hour. If the lunar explorer had any kind of vehicle at all, he could stay well ahead of the terminator, and never see the sun at all, if he didn't want to.

Then, even if it were necessary to remain in the sun side, there would be numerous shadows because of the uneven terrain. Since there is no air on the moon, heat is not carried by moving wind from sunlit areas into the shadow. A lunar shadow is cold, no matter how hot the sun-drenched areas about it may be.

In the equatorial zone, shadows virtually disappear when the sun is high in the air and that, of course, is when the heat is worst and most dangerous. In high latitudes, however, there are always shadows in one direction or another and there are spots within the crater rings, far north or south, that never get sunlight. An exploring base might even be set up under the protection of such a crater ring.

But might not the absence of sunlight be just as bad? Without any oceans to serve as heat reservoirs, without any air to circulate warmth from lighted to unlighted regions, temperature plummets at once as soon as sunlight is withdrawn. During the two-week-long night, temperatures reach something like 250° below zero Fahrenheit just before dawn.

That, however, sounds worse than it is. An explorer encased in his spacesuit in the coldest part of the lunar night is surrounded by vacuum. There is no piercing wind to carry heat away from him and the ground underneath is a very poor conductor, too. He can lose heat only by radiation and that is a slow process. In other words, the explorer is a kind of living Thermos bottle and his own body heat will probably suffice to keep him warm even under the most frigid conditions.

For that matter, if heat or cold is any problem at all, the explorers can always dig underground and set up a base several feet below the surface (see Chapter 31). The lunar surface is so poor a conductor that the broiling heat of day and the frigid cold of night affect only the outermost skin of rock. A little way below the surface, the temperature is unchangeably comfortable.

A base underground would offer at least partial protection against cosmic rays and would also offer protection against the fall of meteorites. The moon, like the earth, is subjected to a constant rain of tiny particles from space, but on the moon there is no atmosphere to burn them into harmless dust.

To be sure, most of the meteorites are so tiny as to be harmless. At most, they might gouge tiny scratches in an explorer's faceplate. However, there would be occasional pieces large enough to penetrate weak points in the suit.

It is possible that explorers might carry umbrellas of thin aluminum to guard against this. Flying grit would expend its energy harmlessly in puncturing the umbrella. Larger pebbles capable of passing through the aluminum without being appreciably slowed could still be fatal, but it would be unreasonable to worry about such a low-probability event. The explorer would be in greater danger of being struck by an automobile every time he crossed a street on earth.

That leaves the matter of the terrain. The usual illustrations of the lunar surface by imaginative artists show it to consist of crags and ravines, of steep rugged mountains and of jagged valleys. This is not so. The moon is gravelly and pebbly in its flat portions (according to the photographs sent back by Lunar IX and Surveyor I) but the mountains and craters represent rather gentle slopes. Since the moon's gravity is only 1/6 that of the earth, an explorer will have no trouble negotiating the terrain even while wearing a bulky and massive spacesuit. If he has a vehicle at his disposal, he is on easy street.

There is some concern that at least some part of the moon's surface may have a coating of dust in fairly thick layer. Close photographs have shown no clear indication of this but the possibility is not entirely removed. If so, the explorer might find he could carry through his exploration only on broad sledlike vehicles—but that can be done, too.

All in all, once we manage to land men on the moon, with adequate equipment and supplies, actual exploration of the moon will probably be considerably less dangerous than the exploration of Antarctica.

But why explore the moon at all? What is there to find? There is no indication that there are any precious substances on the moon. It is probably made of rock similar to that which builds up the earth's crust. Anything common on the moon would be common on the earth, anything rare here would be rare there, too. Even if we found a cache of diamonds on the moon, or a rich strike of uranium, how would we get it back to the earth at anything but prohibitive expense?

However, mankind seeks more than material wealth. There is, first and foremost, knowledge. Only by actually landing on the moon and exploring it can we enrich our knowledge about the moon itself. Nor should you ask what good it is to know about the moon, for the knowledge we gain of the moon may tell us much about the earth and ourselves.

Both earth and moon were formed billions of years ago, it is believed, by certain natural processes. Astronomers are at loggerheads concerning the exact details of those natural processes. There might have been clues built into the structure of the earth but if so, those clues have long since been obliterated by the action of water, wind, and living things.

For instance, the earth must have been subjected to the fall of large meteorites through its history, but there is the clear mark of only one such fall—a depression, like a tiny lunar crater, in Arizona. That crater, only a few thousand years old, is in a desert region where it has been comparatively safe from erosion, and that is the only reason it has survived during its short lifetime. What about older craters? There are faint remnants of some, but nothing that can be studied clearly.

On the moon, however, where the processes of erosion are much slower and less drastic than on earth, all the marks of creation must be present with remarkable freshness. From the moon's surface, we should be able to read the moon's past and this will tell us the earth's history also. We may find out, for the first time, just how planets are created (and perhaps why the moon is so impossibly large).

Then, too, the moon would be an astronomer's paradise. Here on earth, in the latitude of its chief population centers, the night is only 18 hours long at most. The air dims the stars, and temperature variations in the atmosphere cause their light to shake and twinkle. City lights bleach out the stars; fog and clouds obscure them; man-made dust and smog blot them out, Our telescopes must, in desperation, be placed in isolated regions on top of mountains, and still man's habitations encroach.

But the moon—there the nights are two weeks long and there is no air or man-made dust. The stars can be seen steadily and brightly. Better still, the planets can be seen clearly. A small telescope on the moon would see the details of Mars' surface more clearly than even the largest telescope on earth could. We would see Mars better than under any conditions short of a close-flying Mars-probe like Mariner IV.

The sun, too, could be studied with particularly good results from the moon. None of its radiation would be cut off, its corona could easily be made visible at all times.

Could not all such observations be made from a space station, or even from an automatically instrumented satellite? Perhaps, but the moon would support a far larger and more complicated astronomical observatory than a satellite could, and do so in far greater comfort than a space station could.

In addition, there is no substitute for the moon for radio astronomy. It is ony 30 years since astronomers have begun to interpret the radio waves that reach earth from the sky and to deduce many interesting facts from them (see Chapter 19). And already, radio astronomers are concerned that man's own increasing use of radio waves may soon blot out the weak signals from the sky.

A space station would do no good in this respect for earth's "radio racket" would fill the space around it. On the moon, however, an astronomical observatory could be set up on the far side, the one that never sees the earth. With a couple of thousand miles of rock between the observatory and the noisy earth, astronomers could listen to the music of the spheres in complete and blissful silence.

Ten years on the moon could tell us more about the universe than a thousand years on the earth might be able to.

It is all very well for explorers and scientists to have fun on the moon, but it would be nice to feel that there could also be something on the moon for the ordinary person—for you and me.

Suppose trips to the moon became routine; is there any reason for an ordinary earthman to go?

Yes, indeed. There would be the excitement of strange places, the thrill of a completely new kind of surrounding, and the wonder of sights never before seen.

The sun (viewed through special protective devices, or better still, by indirect means such as television) would be a fearsome object, and the incredibly numerous and bright stars of the night sky would be beautiful. Nothing, however, would be as magnificent as the one sight in the lunar sky that cannot be duplicated here on earth. Any tourist would consider the expense and danger of the trip repaid in full once he saw the earth in the sky.

The earth as seen in the moon's sky (going through the same phases the moon goes through for us) is nearly four times as wide as the moon we see here on earth. It is about 13 times the area and, since it reflects much more light than the moon does (thanks to the earth's possession of clouds), the earth in the moon's sky is 70 times as bright as the moon we see here!

Because the moon always turns one of its faces toward the earth, the earth seems to hang motionless in the moon's sky. (As seen from some spots on the moon, it is always directly overhead. From other spots, it is always low in the sky, in some particular direction. And from nearly half the moon's surface, of course, it is never seen.)

Every once in a while the sun, in its passage across the moon's sky, would move behind the earth. (Here on earth we see this situation as a lunar eclipse, an eclipse of the moon.) The sun will stay behind the earth for as long as an hour and the moon's surface will grow dark—but not entirely.

The sunlight will light up the earth's atmosphere all around its globe and this will glow as a bright orange circle around a perfect blackness and cast a ruddy glitter on the moon's surface. Beyond the orange ring in the sky will be the faint white corona of the sun. No one who sees this sight will ever forget it.

On top of all this, there will be the excitement of experiencing low gravity. The sensation of feeling feather light, of being able to jump far and high, will be a great novelty. Of course, control of the body under conditions of low gravity will not be simple and it will be easy to take tumbles. The man who has his "moon-legs" will have ample opportunity to laugh at the tenderfeet who are still in the process of adjustment.

Indeed, there may be individuals who will see the moon as something more than a place for a tourist's visit. They may want to stay.

Once mankind makes a start on the moon, such permanent stays may well become possible. The moon itself can be used as a source of material and energy so that a lunar colony might become largely independent of the earth. Nuclear power stations based on lunar uranium could be used for energy, as could the bright sunlight, never dimmed by clouds. Hydroponic farming, powered by such energy, could supply ample food.

What's more, the moon is not, after all, as dead a world as has been thought. In recent years, signs of volcanic activity have been reported so that there may be internal heat that can be used as an energy source.

Then, too, though there is no air or water on the moon's surface, what about the regions under the surface? It is not completely impossible that traces of air and water linger in crevices under the surface and, if so, these could be salvaged for the use of a lunar colony.

Indeed, some speculate that it may even be possible for primitive microscopic life to have developed in these underground caches of air and water (see Chapter 20).

Even if air and water do not exist underground, the necessary hydrogen and oxygen (and other substances too) could be obtained from the rocks themselves, provided only that energy is available.

The time may come when huge underground caverns may be gouged out below the lunar surface and made airtight. Lunar cities could slowly be built, cities in which men and women can go about in absolute comfort without spacesuits; where children can be born and the generations pass.

Such moon colonists might become so adapted to the moon's weak gravity as to become unable to endure the earth's strong pull. If that were to happen, the colonists would be isolated from the home world. Fearing this, it is likely that the colonists will take care to exercise. Large centrifuges, for instance, can mimic earth gravity, and regular stints within such centrifuges will keep the colonists in tone.

The possibility of the colonization of the moon is a particularly exciting aspect of the future. It is the strong and creative who undertake the dangers of a long migration to a new land. Colonies, stimulated by the hardship of the frontier, often outdo their homeland. The ancient Greeks in Asia Minor and Sicily were more prosperous than those of Greece. The Europeans who built up the United States, Canada, and Australia outdistanced the old continent.

Could it be that a society established on the moon would outdistance us, form a bright new civilization, solve problems with which we struggle vainly, and eventually come back to teach us their new and better ways (as America has, more than once and in different ways, come back to rescue Europe)? It is just this matter that will be discussed in greater detail in Chapter 31.

Chapter 30   The Solar System and the Future

In less than 10 years after the first satellite was placed in orbit about the earth, men have been placed in orbit and remained there for up to two weeks. Some of them have emerged from the space capsule to "walk" in space. Unmanned satellites have made soft landings on the moon, and others have skimmed Venus and Mars to make observations that could not have been made from earth's surface.

What lies ahead of us now? If mankind can advance so far in space in less than 10 years, where will he go in the

First published under the title "How Far Will We Go In Space?" Reproduced from *The World Book Year Book*. Copyright © 1966 by Field Enterprises Educational Corporation.

next 10 years? In 20? In a century? Is there anything we *cannot* do in space by 2100, for instance?

Suppose we begin by asking where we stand on the matter of unmanned exploration of space. There the greatest barrier was overcome in 1959, when, for the first time, a rocket was hurled upward by man at a speed of more than seven miles a second. At such speed, a rocket is not confined by gravity to an orbit about the earth. It "escapes," and goes into orbit around the sun. The faster a rocket is hurled, the larger is its orbit about the sun. If it is made to slow down, it will drop closer to the sun. By carefully adjusting a rocket's speed in mid-flight maneuver, we can place spacecraft close to Venus or Mars, even though these planets at their closest are many millions of miles from us. Mariner II executed a passage within 22,000 miles of Venus in 1962, and Mariner IV passed within 6,000 miles of Mars in 1965.

It would not take much more refinement to plot the course of an unmanned probe to Jupiter, Saturn, and beyond. This is something that could be done now if our space scientists were not committed to other tasks of greater importance.

It is not enough, however, simply to send a piece of metal toward Jupiter. If a planetary probe is to be useful, the ship must send back signals. The signals tell us its position and provide us with other vital information. From how far out in space can we reasonably expect to be able to receive such messages?

Already space scientists have sent radar waves to Jupiter and have (possibly) detected the reflection. The distance of such a round trip to Jupiter is about 800,000,000 miles. This is quite an advance over the time, just after World War II, when it was a great feat to bounce radar waves off the moon—a round-trip distance of less than 500,000 miles. It seems possible that by 1975 or so, our techniques will have developed to the point where we could produce a radar beam that could bounce off a body 4,000,000,000 miles away—the distance to Pluto, which is the most remote planet known in our solar system.

We will soon be in a position, then, to explore the entire solar system with unmanned probes. By the year 2000, we might well have launched one or more probes toward every one of the planets in the solar system. The results of these probes will not, however, all be known by then, for trips to the outer reaches of the solar system take a great deal of time. Mariner IV took more than eight months to reach the vicinity of Mars. If it were traveling to Pluto, many years would be required for the flight.

Can we explore beyond the solar system? After all, if we propel a rocket at a speed of more than 26 miles a second (escape velocity from the sun at our distance from it), it will no longer remain in orbit about the sun. It will leave the solar system forever. If we aim it correctly, it will eventually approach Alpha Centauri, the nearest star to our system, or, for that matter, approach any other object toward which it is aimed.

Unfortunately, though, even the nearest star is almost 7,000 times as far away as Pluto. The flight of an unmanned probe to Alpha Centauri might well take many centuries. Nor does it seem that we will be able to develop communications beams of sufficient power to track a probe all the way to the stars. Certainly we will not, in the next century or so (see Chapter 22).

And what about manned flight? A lunar probe taking pictures of the moon does not compare in excitement with a man landing on the moon. And will reaching the moon be the end? Can we expect human beings to land someday on the surface of Mars or Jupiter? Where can we draw the line and say: "Here man is not likely to go in the next century and a half?"

Man can explore space in four stages: in journeys that last days, months, years, centuries. The first stage, a trip of a few days, will take him to the moon. Mankind hopes to have a man on the moon by 1970. Is there anything to stop us from achieving this goal, other than the possibility of mechanical failure?

There are two hazards that are being thoroughly studied. First, an astronaut would be exposed to weightlessness for as long as a week. Is this dangerous? Well, men have been placed in orbits for two weeks and have survived in good condition—weightless all the while. That seems to take care of that. Secondly, astronauts will be exposed to radiation in the Van Allen belts about the earth and to bursts of high energy particles from the sun, as well as to cosmic rays from beyond the solar system. Can they be protected from this? Dozens of satellites have been sent out by the United States and the Soviet Union to study the nature and effects of radiation. Nothing has been reported so far that would make a lunar flight impossible.

The only obstacle, then, that is keeping us from a flight to the moon right now is the need to work out the engineering details necessary to make it reasonably certain that we will not only send an astronaut there, but also bring him back alive. Once we reach the moon, there seems to be nothing to prevent us from ferrying machines and supplies there to build a permanent base (see Chapter 29).

By 1980 or 1985, such a base may exist. From an astronomical observatory on the moon, knowledge can be gained which can smooth the way for more extensive voyages of exploration. What's more, the moon, with its lesser gravity, could eventually be used as a more economical launching pad for extended voyages than the earth itself would be.

The second stage of space exploration—trips of a few months—will place the inner solar system within our grasp. This includes the planets Mars, Venus, and Mercury. Of these, Mars is the least forbidding. Despite its extremely thin and arid atmosphere, Mars just possibly may have simple life forms on its surface (see Chapter 20).

The main difficulty in reaching Mars involves the length of the journey. Before men can reach Mars, they must spend six months or more in space. Can they remain in isolation that long? Can they carry sufficient supplies? Can they endure weightlessness that long?

Let's consider these problems. Isolation need not have serious effects. Four or five centuries ago, men made voyages that lasted several months across wide oceans under conditions almost as dangerous for them as a flight to Mars would be today. They were even more isolated then than a space traveler would be now. They were truly cut off from home, whereas an astronaut would be in radio communication with the earth at all times—with the encouragement of all humanity constantly in his ear.

The problem of supplies is one for which solutions are

being found. First of all, it will not be necessary to pack aboard a spacecraft to Mars the several tons of water and oxygen each man would need during the trip. Instead, the spaceship would carry a miniature chemical plant which would distill and purify waste water and process carbon dioxide to recover oxygen for breathing. It is not contemplated, however, that food would be produced aboard ship. Food would be brought along in freeze-dried packages.

What about weightlessness? It would seem that a man in a state of weightlessness for six months or more would suffer physical harm. If, however, a specially designed spaceship (or part of one) could be spun steadily, a centrifugal effect would be produced within it that would push the astronaut out toward the walls. This would have the same effect upon him as a gravitational field. It would take no energy to keep the ship spinning once it was put into such motion, and the effect might well be to keep the astronaut healthy and comfortable.

If these problems are solved, astronauts may land on Mars by 1985, and there may be a permanent station there by 1995. Stations might also be established on the two tiny Martian moons, Deimos and Phobos, which have no atmosphere and virtually no gravity.

What about the danger of radiation on these month-long trips? The principal danger would come from high-energy particles emitted at unpredictable intervals from flares on the sun. Although spaceships to Mars would be moving away from the stronger radiation of the sun, radiation shields would have to be provided to protect the astronauts during periods of intense solar activity. Mars itself has no detectable radiation belts to worry about once the spaceship nears the planet.

Trips to Venus and Mercury would take no longer than the trip to Mars, but those to Mercury would take considerably more energy because of the orbital mechanics involved (maneuvering an orbit in the presence of the nearby sun's gigantic gravitational field is difficult).

Neither Venus nor Mercury is expected to have any radiation belts to speak of. Both are, however, in the direction of the sun, whose radiation increases dangerously as it is approached. If the radiation danger can be overcome, and in all probability it will be, Venus and Mercury can be reached before 2000.

Establishing permanent bases there is another matter. The surface temperature of Venus, as measured by Mariner II, is about 800° F. This is the temperature all over the planet's cloud-shrouded surface, both day and night, so it must be at least that hot under the surface. There would be no escaping the heat by burrowing underground. Unmanned probes could reach Venus's surface, and a manned expedition might make a temporary flight beneath the clouds, but it seems unlikely that a permanent base will be established on Venus in the foreseeable future.

Mercury is a better prospect since it has no atmosphere to conserve the heat and spread it over the entire surface. Until very recently, it was thought that Mercury presented only one side to the sun, so that one side was always unbearably hot, while the other was almost at absolute zero. If that were so, we could land on the cold side. It is simple to establish an artificially heated base, whatever the cold. Now, however, we know that Mercury slowly rotates with respect to the sun, so that each part of its surface has a day and a night about 59 earth-days long.

During the night, however, any spot on its airless surface has ample opportunity to cool down. This means that any expedition landing on Mercury would have to do so at a point far enough into the night shadow for the surface to have cooled down. An underground base would then have to be dug before the landing point had circled into sunlight again.

Mercury approaches to within 28,000,000 miles of the sun. Can men ever expect to approach even closer? One possibility exists. There is a tiny asteroid named Icarus, which at times passes within a few million miles of the earth. It has a very flattened orbit. At one end, it reaches halfway to the orbit of Jupiter, but at the other it falls in toward the sun, speeding about it at a distance of only 19,000,000 miles. If an expedition could reach Icarus while it was passing near the earth and implant the proper instruments hastily, marvelous observations could be made of the neighborhood of the sun, the charged particles it emits, and the magnetic field it produces.

Any closer approach to the sun by man than Icarus would seem unlikely. Spaceships, manned or unmanned, could be made to skim about the sun at closer distances, but the heat and radiation would very probably be fatal not only to men but even to instruments, unless they were particularly well protected. It seems doubtful, therefore, that in the next century and a half, men will succeed in doing better than Icarus.

The third stage of space exploration—that which will involve voyages lasting years—will carry us to the vast outer solar system. This can be done in graduated steps. Between the orbits of Mars and Jupiter circle thousands of asteroids. A few of them are a hundred miles or more in diameter. Ceres, the largest, is 480 miles in diameter. Once we get to Mars, we will be able to reach the asteroids without too much additional trouble. Perhaps as early as 2000, man will have landed on Ceres. Step by step, other asteroids may be reached. One of the most interesting is Hidalgo. It has a very elongated orbit. At one end, it approaches to within 24,000,000 miles of the orbit of Mars. At the other end, it recedes as far from the sun as does Saturn. Hidaglo's orbit is quite tilted as compared to the orbits of the various planets, so it comes nowhere near Jupiter and Saturn. Still, if an expedition could land on Hidalgo when it was near Mars, men could remain in space for years, studying conditions in the outer solar system at their leisure, knowing that they would eventually return to the neighborhood of the orbit of Mars.

Astronauts could tackle the outer planets one by one, establishing themselves firmly on one, then progressing to the next one. To make these trips, however, even under the best of conditions, astronauts would have to spend many years in space, if spaceships are equipped with the chemical rockets of the kind used today. Unless a new kind of rocket is developed, it may well be that man will never pass beyond the asteroids.

The use of nuclear rockets is a possibility. Rockets might be driven by a series of atomic explosions or by exhaust gases expelled by the heat of a nuclear reactor. In either case, rocket ships could be kept under acceleration for longer periods, and would attain higher speeds.

Then, too, there is an ion rocket now being developed by scientists. Ordinary rockets achieve their thrust by hurling large quantities of heated gases backward. This brute force is necessary to lift the spacecraft above the atmosphere and push it into an orbit around the earth. Once in orbit, however, and surrounded by a vacuum, a ship might make use of electrically charged atoms (ions) instead. These can be hurled backward by the action of an electric field. The thrust of the ions is very weak, so the rocket's speed increases very slowly. The ion rocket is, however, much more efficient in the long run than an ordinary rocket. Acceleration can be continued for indefinite periods, and speeds approaching that of light itself (186,282 miles per second) could, in theory, be attained. By 2000, when men will have reached Ceres, both nuclear rockets and ion rockets may be in operation. If so, it may be with these that the outer solar system will be explored.

A generation later, say by 2025, we may well have landed on one or another of Jupiter's satellites. A century from now, a landing may have been made within Saturn's satellite system, with plans in the making for reaching the satellites of Uranus and Neptune. By 2100, perhaps men will stand on Pluto, at the very limits of the solar system.

Notice that I mention the satellites of Jupiter, Saturn, Uranus, and Neptune. What about those planets themselves? These four planets are giants with conditions that are far removed from those on the earth. They are frigidly cold and have deep, thick, poisonous atmospheres that have incredible storms and winds of unimaginable violence. Pressures at the bottom of these atmospheres must be thousands of times greater than ours. Nor are we certain as to the kind of solid surfaces they have.

If astronauts ever did reach the solid surface of the outer giants (through the use of a spaceship with some of the properties of the bathyscaphes with which we now explore the oceanic abysses) they would be subject to gravitational pulls much stronger than those which are experienced on earth. These pulls would largely immobilize the astronauts and make the problem of getting off the planet almost insuperable. The difficulties in sending manned expeditions to the surface of the giant planets are so great that for a long time space scientists will be satisfied to send unmanned probes spirally toward Jupiter, Saturn, Uranus, and Neptune. Manned exploration of these planets will not take place in any foreseeable time. But small Pluto can be landed upon.

The fourth stage of space exploration—voyages lasting centuries—will take us to the planets of the nearer stars. As was said previously, the nearest star is almost 7,000 times as far away as Pluto. Why bother?

Well, nowhere in our solar system is there another planet on which man could live comfortably. He would have to live underground or beneath domes (which, however, may turn out to be an exciting forward step in man's progress—see Chapter 31). Nowhere else in the system, outside the earth, can there be anything more than very primitive life forms. Out there among the stars, however, there are sure to be other earthlike planets, which may very likely bear life (see Chapter 22). Some of them might even bear intelligent life. Unfortunately, we cannot be certain a particular planet bears life until spaceships get fairly close to the stars that these planets circle, so that if other life is what we seek, we must explore blindly.

But can other stellar systems be reached?

Certainly the task of reaching even the nearest ones is many times as difficult as that of reaching even the farthest planet of the solar system. A major problem in making such a trip would be to ensure protection against the lethal, high energy particles that would collide with a spaceship, endangering its passengers and instruments. No solution to this problem is yet known. Moreover, even the most advanced rockets we can imagine cannot go faster than the speed of light, and, even at the speed of light, a round trip to the nearest star would take nearly nine years. Round trips to more distant stars would take hundreds of thousands of years.

Even by 2100, when mankind may well be in occupation of Pluto, it seems doubtful that any serious attempt would have been made to send out an expedition to the stars. Does that mean, though, that men will *never* reach the stars? "Never" is a pessimistic word. Scientists have speculated on several means of reaching the stars. The first necessity, of course, is the ability to reach speeds approaching that of light. These may be reached by means of ion rockets or some other technological developments not yet visualized.

Einstein's theory of relativity explains that all internal motions slow down in objects moving at great speeds. Astronauts, therefore, might experience the passage of only a few years in the course of voyages which to men on earth might seem to be lasting hundreds of thousands of years (see Chapter 18). Men could therefore reach even distant stars in the course of their own lifetime, though that would mean saying farewell forever to the earth they left behind.

If it turns out that speeds near that of light are not practical, it may be possible, nonetheless, to live long enough to reach the stars. To achieve this, astronauts could be frozen and put into a kind of suspended animation for decades or generations until their destination was in view. We cannot say as yet, however, whether such suspended animation by low temperature hibernation will ever be practical.

There is a third way out. In place of the small ships used for exploration and colonization of our solar system, a huge ship might be built for voyages to the planets of the stars. Actually it would be a small "planet" itself. On such a "star ship," there might be hundreds, or thousands of men, plus room for agriculture and for herds of animals. Whole generations of men and women might be born, grow old, and die while the star ship traveled from one star to another. The conditions under which such star ship exploration might most likely be practical will be considered in the next chapter.

When expeditions are sent to the stars—by whatever system—we need not expect to see them come back. Even a successful expedition to any but the very nearest stars cannot possibly return to earth in the same century, as we count time. Nor will it be possible to communicate with any human colonies that may be established on the planets of other stars in ordinary fashion. Even if we develop the ability to transmit communication beams intense enough to reach other stars, it will take dozens of years, even centuries, for such beams to reach the colony and an equal amount of time for the colony to answer (see Chapter 22).

Let us summarize then. A reasonable guess is that by 2100, mankind will have explored our entire solar system and will have landed on the surface of any planet, satellite, or asteroid he has tried for, except for Jupiter, Saturn, Uranus, Neptune, and Venus. He will have studied the sun from close range, but not more closely than from a distance of 19,000,000 miles. Mankind will *not* have made any attempt to reach or colonize planets outside our solar system.

After 2100, a long pause may be enforced on mankind. He will probably have gone as far as he can go without developing technical abilities far beyond what he will possess even then. Those space feats which mankind will not have accomplished by 2100 (a landing on the giant planets, a very close approach to the sun, a voyage to the stars) may not actually be impossible, but they are so difficult that mankind may not even attempt them for many centuries after 2100.

Chapter 31   The Universe and the Future

Let me begin by coining an uneuphonious word—*spome* —and defining it.

A spome is any system, substantially closed with respect to matter, that is capable of supporting human life for an indefinitely long period of time.

The earth is a spome and, at present, is the only spome known to exist. Its qualifications for spomehood are obvious. It has supported human life for well over a million years, if we count the hominids generally, and will continue to do

This was presented as a paper to the American Chemical Society on September 13, 1965. Originally published as "There's No Place Like Spome" in *Atmosphere in Space Cabins and Closed Environments,* edited by Karl Kammermeyer. Copyright 1966 by Meredith Publishing Company. Reprinted by permission of Appleton-Century-Crofts, Division of Meredith Publishing Company.

so for the foreseeable future, barring the effects of man's own willful folly.

Furthermore, it is substantially closed with respect to matter. The matter that is added in the form of meteoroid infall or lost in the form of atmospheric leakage is not significant. It does not affect earth's spomic characteristics, nor is it likely to in the foreseeable future.

But a spome cannot be closed with respect to energy.

Life is a process whereby relatively unorganized components of the environment are made more organized. That means that life involves a continuing decrease of entropy and can exist only at the expense of a continuing, and even greater, increase of entropy in the environment generally.

If the earth were closed with respect to energy, mankind, and life generally, would see to it that in a relatively short time, enough oxygen and organic matter would be degraded to carbon dioxide and other wastes to render the earth uninhabitable.

The energy of the sun makes all the difference. It enters the earth system, keeps the atmosphere stirred up and the oceans liquid; it makes the rain fall; and most important, solar energy is utilized by green plants to reconvert carbon dioxide and water into organic substances and free oxygen.

The entropy of the environment, pushed upward by the activities of life, is pushed downward again by the energy of the sun. An equilibrium has been maintained for some billions of years at the expense of the vastly increasing entropy of the sun—which has room for additional entropy increase for some additional billions of years.

Beyond the sun we need not go. For all we know, there are processes that reverse the entropy increase of the sun, and of stars generally, and keep the universe in stable equilibrium forever, as some astronomers have believed (see Chapter 19) but that need not concern us. The sun will endure, substantially in its present form, for some ten billion years and that, on the human scale, is an indefinitely long period of time. Earth may therefore certainly be regarded as a spome.

If the earth were the only spome that could exist, the subject of spomology would be trivial. It would be comprehended by such sciences as geography and geology. But it may be that the earth is merely the only spome that exists so far, and that many others can exist in conception or potentiality. In that case, the subject increases in interest.

It is possible—indeed, it is certain—that elsewhere among the stars (but *not* in our own solar system) there may be other spomes. That is, there may be planets sufficiently like the earth in general characteristics, with a sun sufficiently like our own sun, to serve as habitable planets and therefore as spomes. The figure I have used elsewhere in this book (see Chapter 22) is a possible 640,000,000 in our galaxy alone.

And yet all 640,000,000 lumped together do not in themselves suffice to make spomology a truly interesting study, for they are all merely so many earths. From the broad standpoint of the spomologist, if you see one earthlike planet, you have seen them all. Since we have all indeed seen one earthlike planet, our own, we have seen them all and can forget about them.

What we want, if we are to make spomology interesting, are spomes that are drastically different from the earth. And if we make the subject interesting, we may find—who knows—that it is valuable as well.

Suppose we ask ourselves what makes earth a spome and Jupiter or Mercury nonspomes? If we want to express the difference most succinctly, it is a matter of mass. Jupiter is too massive; Mercury is insufficiently massive. The difference in mass involves, one way or another, almost every quality that goes to make or not make a spome.

If a planet is insufficiently massive it cannot hold either an atmosphere or an ocean of a volatile liquid. If it is too massive, it will hold hydrogen and helium and produce a poisonous atmosphere and, at best, an ammoniated ocean. In neither case, can it be a spome.

If it is very massive, that is probably because it is far distant from its primary and can accumulate matter with little competition from the greater body, and at a temperature low enough to make the dancing molecules of hydrogen (the major component of matter) sufficiently sluggish to be captured. Under such conditions, the planet is too cold to be a spome.

If the planet is insufficiently massive, it is because it is too close to the primary, so that accumulating matter is lost to the greater body and many of the more common elements are, at that distance from the primary, too nimble and elusive to be captured. Alternately, the body is forming too close to a large planet which competes successfully for matter, so that the body itself is a satellite rather than a planet. In the former case, the body is too hot to be a spome, in the latter too cold.

There are exceptions to these rules, of course; known exceptions within our solar system. Our moon seems too large for its place in the system, whereas Pluto seems too small. This departure from regularity leads to theories that the moon is a captured planet and Pluto an escaped satellite.

On the other hand, assuming a sun of the proper type, it is quite reasonable to hope that there is a good chance that a planet of the proper mass would be bound to form at the right distance from that sun and with the proper chemical composition to lead to spomehood.

We might say, then, that the search for a spome is the search for a body of appropriate mass.

But all this is in the course of nature. It works as we are looking for "natural spomes," for spomes ready-made. Let us now add the factor of human intelligence. Only God may make a tree, according to Joyce Kilmer, but perhaps spomes can be made by fools like us. (No, I didn't invent the word in order to be able to make that statement.)

The problem is: Can we make an "artificial spome"? Can we take a body of drastically wrong mass and make a spome of it? In one direction, let's not even try. Bodies too massive to be spomes are quite rare (there are only five in the solar system, counting the sun itself, as compared with many thousands of bodies that are insufficiently massive to serve as natural spomes). The too-massive bodies are, in addition, too dangerous to play with, thanks to their strong gravitational fields and their inevitably enormous atmospheres.

If we look in the direction of bodies insufficiently massive for spomehood, we find at once that the closest body to us, the moon, is an example of the class.

The problem boils down, then, to the conversion of small

bodies into spomes, and the specific version of the problem is, inevitably: Can we make the moon into a spome?

The moon is certainly not a spome now. Thanks to its low mass, it has neither an atmosphere nor free water. But let us consider essentials and not accidentals: An atmosphere can be kept from diffusing out into space by the force of a sufficiently strong gravitational field, but, on a smaller scale, it can be kept from doing so by physical barriers as well.

In other words, we can distinguish two general varieties of spomes: external and internal. An external spome is one with an atmosphere and ocean held to the outer surface of the body by a gravitational field, so that men can live on that outer surface. An internal spome is one with air and water held within an air-tight cavity and with men living on the inner surface. Inevitably, natural spomes are external ones, while artificial spomes must be internal.

Suppose, then, we hollow out a cavity under the moon's surface and supply it with air, water, and the other necessities of life. We might have to begin with capital from the earth, but it is possible that eventually water could be baked out of silicate hydrates in the body of the moon. From such water, oxygen could be formed.

Given a sufficient supply of energy, and a mass of variegated chemical composition such as the moon (or even a much smaller body) the basic chemical requirements can be met on the spot.

Energy is the key, and we are used to thinking of the sun as the energy source. In nature, the only source of energy in quantities large enough to support a natural spome does, in fact, happen to be a star like our sun, but a star—any star—is an incredibly wasteful source. Hardly any of its radiation is stopped by a planet, and only a small fraction of that which is stopped is used. A much smaller source, used with much greater efficiency, will serve the purpose.

A roaring wood fire, whose energy production is a completely contemptible fraction of that of the sun, will warm us in winter at a time when all the sun is insufficient. On the scale that would be sufficient for an internal spome, an ordinary fire is not enough, however. Fortunately, something much better is in sight.

On the large scale of spomehood, only hydrogen fusion

can be relied on as an energy source through an indefinite future. It is large-scale hydrogen fusion that powers the sun, and it may be small-scale hydrogen fusion that will power the earth some day.

I foresee, then—although not in the immediate future—the possibility of the moon being honeycombed immediately below its surface by a growing system of caverns, supplied with all basic materials from the moon itself and with all its energy requirements supplied by fusion power plants. It would be seeded with plant and animal life (and, inevitably, with microscopic life as well) and inhabited by men, women, and children; families who may know no other life, and want none.

The advantages are obvious. The moon will have a controlled environment designed specifically for man; man will have what he wants and needs (in many vital respects) and not merely what he can get. What's more, he will have the advantage of a fresh start. As the United States managed to prosper and flourish partly because it was freed of many of the choking traditions of Europe's bitter past, so the moon, it may be hoped, will be freed of the incubus of earth's past mistakes.

Some disadvantages are also obvious. However confidently we rely on scientific and technological advance, it seems certain that we can never do anything to alter the moon's gravity. The inhabitants of the moon will always be under a gravitational pull only one-sixth that of the earth.

Undoubtedly, they can get used to it, and people born on the moon, knowing no other, will consider such a gravitational force natural. Will men suffer as a result, however, particularly in the transitional period when they may be shuttling between the earth and the moon? Will muscles weakened and bones softened under the influence of lower gravity be able to withstand a return to earth?

The problem might not arise in fullest intensity. Men on the moon could keep in condition with exercise or in centrifuges. Perhaps only a few specialists would need to condition themselves for possible trips to earth, whereas the general population of the moon would find it no hardship at all to remain away from the earth permanently.

Another disadvantage is that an internal spome is liable to accidental catastrophes of a sort to which external spomes are immune. An atmosphere and ocean held to the surface by gravity are absolutely secure. Barring catastrophe on an astronomic scale, nothing can alter the gravitational force and nothing can cause the atmosphere and ocean of an external spome to be lost.

On an internal spome, on the other hand, a cavern punctured by a large meteorite, or ruptured by a landslide, loses its air at once and its water more slowly. Nevertheless, it is to be expected that men will be ingenious enough to minimize the chances of such catastrophes. Furthermore, the cavern of an internal spome will undoubtedly be compartmentalized so that a local catastrophe can be confined to its immediate neighborhood.

Nor is catastrophe in itself a bar to spomehood. There are catastrophes on earth, too. We suffer periodically from the effects of hurricanes, blizzards, tornadoes, floods, and drought, to none of which the moon would be subject. A patriotic moonman might well argue that it was the earth rather than the moon that fell short of ideal spomehood through catastrophe.

But what about the psychological difficulties? Can men really learn to live for extended periods in what is essentially, after all, a cavern? Can he bear to be born and to die there? The answer, in my opinion, is the heartiest possible affirmative. If the cavern is large and comfortable, why not?

It is a mistake to underestimate the flexibility of mankind. Man has already demonstrated abilities to make enormous adjustments. A city such as New York represents, in a way, almost as artificial a spome, one almost as divorced from man's original environment, as the moon would be. Yet man has made the transition from hut to skyscraper over an insignificant period of time. Indeed, a peasant immigrant can adjust adequately to New York in his own lifetime.

Why should we imagine a moonman would be horrified at being "cooped up"? I think it would be much more likely that he would think with horror of a world like the earth, there men had to cling precariously to an outer surface, exposed to the vagaries of an unpredictable and changeable climate. A moonman might no more want to live on earth than a New Yorker would want to live in a cave.

Of course, in thinking of an internal spome, we must fight our prejudices. It is easy to fall into the trap of thinking, vaguely, that an external spome is "natural" and an internal spome "artificial" and that what is natural is good and what is artificial is bad.

The argument might even be advanced that a "true" spome can only be one in which life could develop spontaneously out of nonliving matter, as it did on earth (see Chapter 9). A world that had to be engineered and seeded by a species that already had two to three billion years of evolution behind it might seem no true spome at all, but one that was only able to imitate spomehood through an initially parasitic dependence on a true spome.

But if that argument is advanced, where does Homo sapiens stand? Life did not develop on dry land. The only portion of the earth that is a "natural" spome, in the sense that life arose there spontaneously from simple chemicals, is the ocean. It was only little by little that certain types of living things emerged onto the dry land, a habitat as hostile to the creatures of the sea then as the moon seems to us now.

Some fishy philosopher, if we can imgaine one, might well have shaken his head at the foolish creatures who chose to emerge on land. It would seem a bad exchange to move from the equable environment of the ocean to the violent extremes of the open air; from a plenitude of water to the perennial threat of dessication; from a gravitation-free three-dimensional world to a gravity-ridden two-dimensional world.

Nor are these dangers unrealistic ones, or these disadvantages of the dry land imaginary. Life first invaded the land some 425,000,000 years ago, yet even today, the ocean remains much richer in life than dry land is, area for area. Land animals had to evolve for millions of years before they could develop limbs strong enough to lift them clear of the ground and make both size and rapid movement simultaneously possible. It was some two hundred million years before creatures evolved who possessed internal thermostats and external insulation so that the equable temperature of the ocean might be imperfectly restored. Man himself rose to his hind feet a million and a half years ago and still pays his respects to gravity with flat feet, slipped disks, sinus trouble, potbellies, and numerous other ailments. And to this day he must live in dread of falling, a dread we are usually unaware of only because we are so accustomed to it.

No, no, if we are going to sneer at the moon as an unnatural habitat, we must sneer with precisely equal intensity at the continents of the earth. We live on a portion of the earth artificially seeded from the truly spomic portion; and despite everything, land life remains less rich and, in some respects and by some criteria, less comfortable and less successful than ocean life.

Yet need we be sorry that our ancestors emerged from sea to land? With all land's dangers and discomforts, it opened the way to advances not possible in the sea. In hindsight, we can see that the ocean was a dead end, whereas land offered a new and brighter horizon.

Nor are we being parochial when we argue in this way. Air is far less viscous than water. In water, a creature must either travel slowly or it must be streamlined. The most highly developed sea creatures, the squids, sharks, and fish, are highly streamlined. The land creatures that return to the sea are streamlined in proportion to the extent to which they have returned, if you think of the otter, penguin, seal, sea cow, and finally, the whale.

A streamlined body implies short, stubby appendages, if any, with an exception for the squid's highly specialized tentacles. In low viscosity air, on the other hand, it is possible to be fast-moving and irregularly shaped at the same time, so that land animals can have elaborate appendages. It is to this that man owes his priceless hands.

Consider how, were the porpoise indeed as intelligent as man, the lack of hands would hamper the exhibiting of that intelligence! If we ever learn to communicate with porpoises we may find ourselves with fluked philosophers on our hands; introverts who can think but not do.

Then, too, one can deal with fire only in air and never in water. Only a land creature, therefore, could conceivably develop the technology that begins with the discovery of fire. It is certainly possible to argue that man's advancing technology is not an unalloyed good, but I doubt that even the most inverterate yearner after the good old days before the building of Blake's "dark, satanic mills" could possibly wish to retreat to the days before the discovery of how to start and use a fire.

To use a chemical analogy, the passage from sea to land involved a "phase change" in the progress of life; on that, most or even all of us cannot help but consider desirable.

Is it possible, then, that the passage from an external "natural" spome, to an internal "artificial" spome might likewise involve a desirable phase change? I hate to undertake the role of prophet here; foresight in such matters is as difficult as hindsight is easy. Nevertheless, I will try.

It seems to me, for instance, that however difficult the initial passage from an external spome to an internal one, the end would be a partial cancellation of the difficulties introduced by the previous great life-adventure. In an internal spome, man would return to the equable environment and lower gravity of the sea, without abandoning the low-viscosity environment of the air. An internal spome would have, after a fashion, the best of both land and sea and the worst of neither. Surely something great may come of that.

If we begin with an internal spome on the moon, victory and success there can only inspire attempts at expansion, at forming spomes out of other medium-sized bodies such as Mars and the larger satellites of Jupiter. In particular, though, there may be a movement to internal spomes on smaller and smaller bodies—that is, on the asteroids that exist by the thousand in the space between the orbits of Mars and Jupiter.

Why the asteroids?

Well, consider the matter of efficiency. With the best will in the world, and with all the technological advances likely in the foreseeable future, it would seem that mankind could not burrow very deeply into the skin of the earth, or into the skin of even a smaller body such as Mars or the moon. We may sink narrow bores to the mantle in time to come, but if we are thinking of internal spomes, of large, comfortable and well-appointed caverns, the outer couple of miles is the most that we may consider. (Earth's internal heat, perhaps

270

that of Mars and the moon too, would make deeper caverns uncomfortable anyway.)

This means that virtually all the volume of a planet is unused and serves the men of the spome only by supplying them with the source of a gravitational field.

The asteroids, however, can be spomified completely. They can be riddled and honeycombed. They have no internal heat for discomfort and no significant gravity to make more difficult the shifting of mass. Nor need the caverns be buttressed more than minimally to counter possible collapse. If we except the very largest, *all* of an asteroid can be used. (A nickel-iron asteroid might be difficult to work with, and its composition might not be suitable as a source of raw material for anything except the ferrous metals but, judging by the ratio of iron meteorites to stony ones, we can hope that less than 10 percent of the asteroids will be metallic.)

Nor need an asteroid be considered too small to make an ample spome. Some years ago, I wrote a story about such an asteroidal spome, in which an earthman visiting the asteroid expressed surprise that the inhabitants had room to grow tobacco. His guide to the asteroid replied:

"We are not a small world, Dr. Lamorak; you judge us by two-dimensional standards. The surface area of Elsevere [the asteroid] is only three-fourths that of the State of New York, but that's irrelevant. Remember, we can occupy, if we wish, the entire interior of Elsevere. A sphere of 50 miles radius has a volume of well over half a million cubic miles. If all of Elsevere were occupied by levels of 50 feet apart, the total surface area within the planetoid would be 56,000,000 square miles and that is equal to the total land area of earth. And none of these square miles, doctor, would be unproductive.

In the story I deliberately dismissed one serious problem that would inevitably arise on an asteroidal spome in order that I might concentrate on the sociological point I was trying to make. I avoided any consideration of the fact that the gravitational field on an asteroid is negligible by supplying my storybook spome with artificial gravity.

271

In real life, as opposed to science fiction, an artificial gravity field cannot be set up merely with a wave of the typewriter. One conceivable possibility would be to set the asteroidal spome into rapid rotation. The centrifugal effect would be analogous to a gravitational field directed outward in every direction from the axis of rotation, with some important side effects. The gravitational field so set up would vary markedly with distance from the axis and there would be very noticeable Coriolis effects. The smaller the spome, the greater the angular velocity required for a given maximum centrifugal effect and the more pronounced the variations in the effect and in the obtrusiveness of Coriolis effects.

It seems to me that spinning the spome would not be worth the energy expended and the problems produced. Why not, instead, accept null gravity as a condition of life? Life has, in the past, switched from the essential null gravity of the oceans to the gravity slavery of the land and survived. Why not the switch back?

To be sure, the switch from null-g to g was made over eons of time, and the bodies of the creatures making the switch had to undergo elaborate and glacially slow changes through the force of natural selection. Mankind obviously lacks the time to proceed in this fashion.

But it is not in space science and engineering alone that mankind is experiencing great advances in technology. Biology is undergoing its own revolutionary breakthroughs. It is reasonable to hope that by the time man reaches the point where he can reach the asteroids with a supply of energy sufficient to set up a spome, he will also have learned enough about genetics to engage in meaningful tissue engineering (see Chapter 9). Why may we not suppose that the changes necessary to fit a human body for null gravity can be guided by intelligence rather than left to the colossal blindness of a nature that knows only random change?

A null-gravity body may well be designed differently from our own, but not necessarily radically so. Bones and muscles may be smaller and legs shorter, but I would guess that this would not go to extremes. To whatever extent weight may disappear, the body will still have to handle inertial mass, which would be the same on an asteroid as on the earth.

A null-gravity body would, it seems to me, become utterly graceful in its maneuverings, gaining some of the three-dimensional skills of the fish and birds. We will have a human species capable of flight without having to sacrifice the infinitely useful hand for the sake of a wing.

Land animals might require similar adaptations but, except perhaps for pets, dwellers on the asteroidal spomes could do without them. Plants could be grown at null-g without much trouble. Fish could still be cultivated. Algae culture and the chemical industry might combine to produce food items with the taste and texture of meat if that were desired.

To be sure, a null-g man could never come to earth, or even visit a world as small as the moon, but that should be no more a hardship to him than the fact that we can no longer breathe under water is to us (except when we are drowning).

If we concentrate on this state of affairs, it would seem that there would be two species of man, g and null-g. We are g, of course, as would be the colonists on such large spomes as Mars, the moon, the large satellites of Jupiter, and so on. The inhabitants of the asteroidal spomes would be null-g.

It would not be so much merely the passage from external spome to internal spome that would represent the second phase-change of evolution, as the passage from g to null-g. Might it not be that the future will belong to the null-g? That we g's of earth will now reach a dead end, while the null-g's of the asteroids will find a new and glorious horizon opening up for them? They may advance, leaving the discarded things of earth behind them while we, no more able to follow them than a fish could us, remain as oblivious as fish to their greater glories.

Consider—

First, the null-g species may well outnumber us as time goes on. Honeycombed asteroids may support a larger population, all taken together, than the mere outer skin of the large spomes inhabited by g's. The fact that null-g might be smaller in body (though not in brain) would serve to make their possible numbers still greater.

Second, the nature of the null-g environment will make it certain that they will far outstrip us in variability and versatility. The g people will exist as one large glob (earth's population) with small offshoots on Mars, the moon and elsewhere, but the null-g's will be divided among a thousand or more worlds.

The situation will resemble that which once contrasted the Roman civilization with the Greek. The Romans wrought tremendous feats in law and government, in architecture and engineering, in military offense and defense. There was, however, something large, heavy, and inflexible about Roman civilization; it was Rome, wherever it was.

The Greeks, on the other hand, reaching far lesser material heights, had a life and verve in their culture that attracts us even today, across a time lapse of 2,500 years. No other culture ever had the spark of that of the Greeks, and part of the reason was that there was no Greece, really, only a thousand Greek city-states, each with its own government, its own customs, its own form of living, loving, worshiping, and dying. As we look back on the days of Greece, the brilliance of Athens tends to drown out the rest, but each town had something of its own to contribute. The endless variety that resulted gave Greece a glory that nothing before or since has been able to match; certainly not our own civilization of humanity-en-masse.

The null-g's may be the Greeks all over again. A thousand worlds, all with a common history and background, and each with its own way of developing and expressing that history and background. The richness of life represented by all the different null-g worlds may far surpass what is developed, by that time, on an earth rendered smaller and more uniform than ever by technological advance.

A third difference, and the really crucial one, in my opinion, can best be explained if I now turn to the subject of spaceships.

In the light of what I have already said, we can see that a spaceship is not exactly a spome for a spome must be capable of supporting human life indefinitely. It is rather a "spomoid," something that is capable of serving a spomelike function temporarily.

Spomoids have performed notably well on a number of

occasions already and have at this writing supported two men in reasonable comfort for as long as two weeks.

It is the obvious intention of the human race to explore the solar system by means of spomoids even before any extraterrestrial spomes are established; and, in fact, even if it turns out that the establishment of extraterrestrial spomes is unfeasible. By stages, we might even reach Pluto (see Chapter 30).

But there we would have to come to a halt. Beyond Pluto lie the stars, and the distances there involved are so enormous that the techniques that will have sufficed for the solar system will be completely useless to meet the new situation.

To reach even the nearer stars will involve one of three alternatives:

(1) Straightforward flight from here to the nearer stars and back, the time required being anywhere from a generation to a century or more.

(2) Flight at velocities near that of light, thus introducing a time dilatation effect (see Chapter 18) so that the duration of the flight will seem to the astronaut to be no more than a few months or years. In that case, however, on returning to earth, he will find that the time lapse here has been anywhere from a generation to a century or more.

(3) Flight with astronauts frozen into suspended animation, the effect being the same as in Case 2.

None of these alternatives is pleasing. The astronaut will either have to expose himself to the perils and uncertainties of freezing over long periods of time, or be willing to expend the energies required to reach extremely high velocities. It may well prove that freezing for decades is unfeasible and that the energy demands for time dilatation are prohibitive. If Alternative 1 is chosen as the simplest, the astronaut must not only spend most or all his life on the star ship; he may also have to be prepared to bring up children and grandchildren who will in turn have to take over the star ship and spend their lives on it.

As for those who wait on earth, there are no alternatives. A star ship leaving for a neighboring star may not get back for a hundred years. The original astronauts may shorten the time for themselves by time dilatation, or by freezing, and return scarcely aged, but that does not affect the observers at home. The star ship will still not have returned for a century, and no one in the crowd that waves good-by will be in the crowd that waves hello.

Under the circumstances, stellar exploration would never be a popular exercise for anyone, either on the ship or at home. A few expeditions may set off as *tours de force,* but earthmen, unable to follow them, unable to see the results in their own lifetimes, will lose interest.

But let's consider under what conditions such voyages might become popular.

The longer the exploring trip within the solar system, the more elaborate the spomoid will have to be. By the time the outermost planets are reached, space voyages will have become years in length and a spomoid capable of supporting a crew for years will, of necessity, have a recycling mechanism that would require little further sophistication to serve a crew indefinitely.

The trend in space exploration, then, will be from the spomoid to the spome and, certainly, where stellar exploration is concerned, nothing less than an elaborate spome will be required.

Not only is a star ship a spome, but it is an internal spome, and one of an extreme type. In assembling a crew for a star ship, we are asking earthmen and women to make the transfer from an external spome to an extremely internal one and we may be asking too much.

To be sure, I have been talking about the establishment of spomes all through this chapter—*but by stages!* The change from the external spome of the earth to an internal spome on the moon is, in many ways, a mild one. There is still the chance of communication with earth, there is still the sight of the earth in the sky, even if only on a television set within the cavern, and, finally, the possibility of returning to earth some day.

It is then the men of the moon, accustomed to a mild internal spome, who will go on to spomify Mars and Ganymede. And it will be the far distant colonists, further divorced from the earth by the mere fact that it is not forever hanging in the sky like a large balloon, who will make the further step to the asteroids and the null-g phase-change.

Little by little the inhabitants of spomes would get over any longing for blue skies, open air, the stretch of ocean, the intricate world of mountains, rivers, and animals.

But even a colonist from the moon or Mars would not feel at home on a star ship, which would be null-g, unless it were rapidly rotated—with all the problems that would introduce.

No, the proper crew for a star ship would be null-g people, and there would be no need to recruit them, for an asteroidal spome would be a star ship in itself. Working upward from a primitive spaceship and downwards from the earth, we meet in the middle at the equation: asteroidal spome = star ship.

Under such conditions, a voyage to the stars could be made without hardships whatever. If an asteroid were fitted with rocket motors and made to veer out of its course and away from the sun (the escape velocity from the sun is considerably less in the asteroid belt than it is in earth's vicinity), what would it matter of the null-g inhabitants of the asteroid?

They had always been in a null-g internal spome, and they would still be in a null-g internal spome. They wouldn't be leaving home; they would be taking home with them. What matter how long the trip to a star? How many generations lived and died? There would be no change in their way of life.

To be sure, they would be leaving the sun, but what of that? A dweller of the asteroids would not depend on the sun for anything. Properly space-suited, he might emerge from the asteroid and observe the sun as a tiny, glowing marble in the sky, but nothing more. He may miss that sight and idealize "the sun of home," but such idealizations will evoke nothing more than a nostalgic thought, like the modern city dweller's occasional sigh for the "old home town."

The star ship turning out of its orbit might simply be taking the third and final step in the weaning of life. Once life forms were weaned from the ocean. With the establishment of extraterrestrial spomes, life forms would have been weaned from the earth. With the star ships, they would be weaned from the solar system.

But why should the asteroids bother to become star ships? What do they gain? A number of things:

First, the satisfaction of curiosity—the basic, itching desire to know. Why not see what the universe looks like? What's out there anyway?

Second, the desire for freedom—why circle the sun uselessly forever, when you can take your place as an independent portion of the universe, bound to no star?

Third, the usefulness of knowledge—since a trip of this sort is bound to add to the information possessed and this new information will surely be applied to the problem of adding to the security and comfort of the spome.

Nor need such a journey be dull and uneventful. True, it may take hundreds or even thousands of years to reach a star, and generations may live without seeing one at close quarters, but does this mean there is nothing at all to see?

I can't really guess what phenomena would await the ship and what beauties of nature they will find to admire. One thing seems certain, however: the universe must be better populated than would appear.

We see the stars because they advertise themselves so brilliantly; but small stars are far more numerous than large ones, and dim stars far more numerous than bright ones. Surely bodies that are so small and dim that they can't be seen, except at close quarters indeed, are the most numerous of all.

Perhaps no generation will pass without some dark world coming into view, some material body the star ship may pause to investigate. If the body is large, the star ship couldn't land, but it could still fly by, take up a temporary orbit, observe, and nose it out. If the body were small enough to have a negligible gravity, it could be mined and made to serve as a source of minerals to replace the small inevitable losses suffered by any spome, however efficient the cycling.

When the neighborhood of a star is reached, with its lighted planets, observations might be particularly intense and particularly interesting. The system may contain external spomes: earthlike planets bearing life—even, perhaps, intelligent life.

What a rare phenomenon that would be in terms of human

lifetimes! How fortunate the generation granted such a sight!

Silently, they would observe, watch, and eventually, pass on as the unbearably attractive lure of open space beckoned —and back on the inhabited planet, creatures might talk excitedly of flying saucers— No! I am not advancing this as a serious explanation of the reports of flying saucers here on earth (see Chapter 24).

The neighborhood of a star might offer a chance for refueling, too. I can conceive that the deuterium supplies needed for the fusion reactors might be picked up in the space the ship passes through but such deuterium is spread out incredibly thinly. It would be more concentrated within a stellar system. The neighborhood of a star might then be not only a means of seeing a rare sight, but also a chance to stock up on deuterium—enough to last another million years or so.

If an asteroidal belt were encountered about some star, a landfall might, in a sense, be made. The star ship could take up some appropriate orbit. Other asteroids could then be made into spomes. The colony would divide and new ones would be set up. Eventually one or more of them—or all of them—would set off as star ships themselves. Perhaps an old, old star ship, worn past the worthwhileness of repair, can be abandoned on such occasions—undoubtedly with much more trauma than ever the sun and earth were abandoned.

In fact, there might almost be an "alternation of generations" over the eons as far as the star ships were concerned. There would be a motile generation in which the star ships moved steadily across the vastness of space but in which population increase would have to be tightly controlled. There would then be a sessile generation after an asteroid belt was encountered, when for a long period of time there would be no motion, but the population would proliferate.

With the conclusion of each sessile generation, there would be a proliferation of star ships. As the years passed and lengthened into the hundreds of millennia, the star ships would begin to swarm over the universe—all of it their home.

And every once in a while, perhaps, two spomes would meet by arrangement.

That, I imagine, would involve a ritual of incomparable importance. There would be no flash-by with a hail and farewell. The spomes, having contacted each other in a deliberate search over vast distances, would be brought to a stand relative to each other and preparations would be made for a long stay.

Each would have compiled its own records, which it could make available to the other. There would be descriptions by each of sectors of space never visited by the other. New theories and novel interpretations of old ones would be expounded. Literature and works of art could be exchanged, differences in custom explained.

Most of all there would be the opportunity for a cross-flow of genes. An exchange of population (either temporary or permanent) might be an inevitable accomplishment of any such meeting.

And yet it may happen that such cross-flows will become impossible in an increasing number of cases. Long isolation may allow the development of varieties that may no longer be interfertile. The meeting of spomes will have to endure long enough, certainly, for a check on whether the two populations are compatible. If not, intellectual cross-fertilization will have been carried on, at any rate.

Eventually, perhaps, space will carry a load of innumerable varieties of null-g intelligences, all alike in that space is their home (and, indeed, "space-home" is what the shortened "spome" stands for); in that they are intelligent; and in that they are descended from the inhabitants of some planet that may no longer exist in their memory even as a component of legend, and from which the initial load of humanity may long since have vanished.

It may even be that Homo sapiens will not be the only species to make the transition to a star ship culture. Perhaps there is a crucial point, reached by every intelligence, from which two roads branch off, one leading to the true conquest of space and the other to a slow withering on the planetary vine.

Out there, perhaps, are many creatures waiting for man

to join them. And when we do, we may find ourselves united with them not in terms of material body resemblances, but in the life we lead and in the intellect we cultivate.

Is this, then, the consequence of the new phase change that will make space exploration truly possible? Or am I only stumbling in a vain attempt to see the unseeable? Perhaps the essential point of the phase change is as far beyond my grasp as the smell of a rose is beyond the grasp of a fish or a Beethoven symphony beyond the grasp of a chimpanzee.

But I tried!

PART III

CONCERNING SCIENCE FICTION

Chapter 32   Escape Into Reality

Of all the branches of literature, science fiction is the most modern. It is the one literary response to the problems peculiar to our own day and no other.

The literature of the main stream is, at its best, virtually timeless. It deals with the tensions within the human mind and soul, and with the interrelations of man and man. Presumably, while human biochemistry and psychology remain essentially unaltered, penetrating studies of this nature will keep their value over the generations. Certainly, Homer and Shakespeare show no signs of decay. It is not with the main stream that I intend to deal.

In the more time-bound realm of the specialized literatures, the writer finds his inspiration in a more or less stylized world of the present or past. The mystery, the sport story, the adventure story, the romance, are all played off against a contemporary background familiar to the reader. The historical novel and the western are set against patterns of the past which are somewhat less familiar but can be quickly accepted.

In each case, the background is "true." It may be dismissed as possessing little intrinsic value for that reason; as being of importance only as the setting against which the particular human drama is performed. It has all the unimportance of the painted backdrop in the theater or the property armchair set in place so that a character may reach it in a fixed number of steps designed to fit smoothly into the action of the play.

In a completely different class fall those examples of specialized literature in which the background or setting has as little relation to reality as do the characters themselves. Less, sometimes. In such literature (so used are we to the tameness and good behavior of the background and to the fictitiousness only of the characters) there is actually a tendency to let the setting itself assume first importance. It is this which gives this kind of literature a completely different "feel" from the more usual kind.

There are three main types of such "false-background literatures" which, in order of decreasing age, are (1) fantasy, (2) social satire, and (3) science fiction.

Fantasy is probably as old as speech. In a primitive world, where most of the aspects of nature and of conscious life were unknown and apparently unknowable except by direct revelation, man's attempts at explanation led straight to fantasy.

A dream of a person already dead would give rise to stories of ghosts. The ruinous effects of storm and drought would serve as inspiration for tales of malevolent spirits. Dimly known facts would distort to wonders, so that rhinoceri became unicorns, sea cows became mermaids, and skulls of prehistoric Sicilian elephants became one-eyed giant cannibals.

For that matter, was fantasy really fantasy until the dawn of our own sophisticated age? Is a ghost story a fantasy to one who believes firmly in ghosts? The background, which to ourselves seems to have no relation to truth, *was* the true background to our ancestors. In that respect, fantasy before our own age was simply another aspect of literature against a familiar background.

To be sure, modern fantasies are written against a background now known to be unrelated to reality and are read by readers who are thoroughly aware of it. Yet the neo-fantasy still finds its inspiration in the deductions of the past. The tales still deal with ghosts and vampires, with witches and demons, with the uses of charms, and the dangers of the devil. Such stories, nowadays, are most successful when written lightly and with the intent only to entertain. They no longer frighten.

Social satire is, on all counts, more sophisticated than fantasy. Whereas fantasy is a universal type of folk literature, social satire is the work of an advanced intellect trapped in a society that does not welcome criticism. (One might almost say, simply, "trapped in a society" without the qualifying clause, for what society welcomes criticism?)

In its earliest form, social satire found shape in animal fables like those made famous by Aesop. In such fables, talking animals, with a human society imposed upon their animal characteristics, behave so as to expose man's follies and crimes. The audience laughs and nods in agreement, not at all annoyed at finding animals foolish and criminal, enjoying their own superiority, in fact.

It is on the aftertaste that the satirist depends—the second thoughts later on that after all there is an application close to home in that fable. And because the listener has been seduced into accepting the moral of the tale, since it points the finger of disapproval upon an animal and not upon himself, he is less able now to cast it off.

The parables in the Bible and the funny stories told by Lincoln were designed to make their points indirectly and slowly, and to drive them the deeper for that.

Social satire graduated from the anecdote to the treatise, and the most famous example is Sir Thomas More's *Utopia*. That book deals with the society of a fictitious island. (The word "utopia" means "no place" in Greek, just as Samuel Butler's similar *Erewhon* is "nowhere" spelled almost backward.) Thomas More uses his fictitious society as a whip upon the back of his own society. Utopia is praised as just and virtuous for those aspects of it which were most conspicuously lacking in More's own society. The reader could not but agree with More that here indeed was an ideal society. Then later, slowly, the reader finds himself dissatisfied with his own world just because it is not a Utopia.

Gulliver's Travels by Jonathan Swift is an example of a book that satirizes in both styles. The Lilliputians of the first book and the Laputans of the third are ridiculed for follies (drawn to excess) common to the society of Swift's day. The Brobdingnagians of the second book and the Houyhnhnms of the fourth are praised for those virtues which Swift's own society conspicuously did *not* possess.

It is possible to mistake social satire for science fiction simply because a society different from the real one is described. It is particularly easy to do this because occasionally in their description of fictitious societies, satirists may include details of a science or technology more advanced than their own. For instance, in *Utopia*, More describes the use of incubators to raise chicks; and in the third book of *Gulliver's Travels*, Swift describes a fictitious discovery of two fictitious moons of Mars (all of which later turned out to be coincidentally correct in amazing detail).

Nevertheless, it is important to realize that social satirists were not primarily interested in their fictitious societies as such. The satirists kept their eyes fixed firmly upon their own societies and used the creations of their imagination to point moral lessons. Their fictitious societies were not what *might* be, but only what *should* be or *should not* be.

In the last century, social satirists have deliberately turned to scientific advance as a tool in their trade. There have been Edward Bellamy's *Looking Backward*, Aldous Huxley's *Brave New World*, and George Orwell's *1984*, to name the best-known.

It is almost inevitable that these be considered as science fiction, and yet they are not primarily so. The author's intent is entirely moral. Bellamy praises his society and Huxley and Orwell denounce theirs, each with the desire to work some change, by this praise or blame, upon matters they find deplorable in their own society.

It is social satire still, for all its science.

What then *is* science fiction?

Science fiction, like fantasy and social satire, deals with a background that is not "real." Unlike fantasy, however, its backgrounds are not completely unrelated to reality, but represent a more or less plausible extrapolation of reality. Unlike social satire, the unreal background is dealt with for its own sake, not for its moral application.

Science fiction may be defined as that branch of literature which deals with the response of human beings to advances in science and technology.

Actual change in science and technology, occurring quickly enough and striking deeply enough to affect a human being within the course of his normal lifetime, is a phenomenon peculiar to the world only since the Industrial Revolution (with some temporary and local exceptions). That is, it is a phenomenon that has existed in England and the Netherlands since 1750; in the United States and Western Europe since 1850; and in the world generally since 1920.

The first well-known writer who responded to this new factor in human affairs by dealing regularly with science fiction, by studying the effect of additional scientific advance upon mankind without placing primary emphasis on moral judgments, was Jules Verne. In the English language, the early master was H. G. Wells. Between them, they laid the groundwork for every theme upon which science fiction writers have been ringing variations ever since.

It was not until 1926 that a special market was set up intended exclusively for the products of the science fiction writer. It was in that year that Hugo Gernsback first published *Amazing Stories*. By 1930, three other science fiction magazines were on the newsstands.

Slowly, it became possible (economically) for a young man to decide to make a career out of science fiction writing, but it took ten years for enough writers to be developed to enable the field to attain maturity.

The period of mature science fiction is dated most frequently from the moment when John W. Campbell, Jr. took over the editorship of *Astounding Stories* (which he quickly retitled *Astounding Science Fiction*). That was on October 6, 1937.

To Campbell, science fiction was essentially as I have defined it above. He turned the emphasis of the science fiction story from that of adventures with new inventions or adventures on other worlds (a kind of super-western with spaceships replacing horses and ray guns replacing revolvers) and made it into an increasingly mature consideration of possible societies of the future.

After the dropping of the atomic bomb, a new hindsight respectability fell upon science fiction. Many who had thought stories about atomic warfare (printed in reasonably accurate detail as early as 1941) ridiculous—or even pathological—revised their thoughts hurriedly. The audience increased. The mass magazines began to publish occasional science fiction. Book publishers (notably Doubleday and Company) began to put out lines of science fiction novels. New magazines were published.

By 1950, *The Magazine of Fantasy and Science Fiction*

and *Galaxy Science Fiction* appeared and these, together with *Astounding* (now renamed *Analog Science Fact—Science Fiction*), are commonly considered the "Big Three" of the field.

The editorial policies of the "Big Three" offer an interesting contrast. All publish science fiction, but *Analog* adheres most rigidly to science fiction in the pure sense, as here defined. As the name implies *Fantasy and Science Fiction* adds a generous helping of modern fantasy, while *Galaxy* adds as generous a helping of social satire. In this way, each of the three major branches of "false background literature" is represented.

Many people (including even some science fiction readers) may place no importance upon science fiction at all—except perhaps as a means of affording some amusement to an occasional reader.

This represents a serious underestimate of the importance of the field.

The underestimate exists in part as a result of the fact that the forms of "science fiction" most familiar to the general public are the comic strip adventures of individuals such as Flash Gordon and Superman, and the Hollywood output dealing with the various types of monsters.*

Neither comic-strip nor the usual Hollywood version is really science fiction. Therein lies the confusion. Rather, both are the result of adding a thin veneer of scientific-sounding mumbo-jumbo to a very old type of literature, the adventure-fantasy. Substitute for the dragon that is slain by Siegfried the equally fabulous monster slain by Flash Gordon, and there are few other changes of any consequence that need be made. The Chimera that devastates the countryside

---

* Since this chapter was first written, the movies and TV have done some worthwhile things. As I write, the movie *Fantastic Voyage* has just been released and this represents an imaginary trip through the human bloodstream by a ship and crew miniaturized to the size of a bacterium—a movie that spared no expense on special effects and made considerable effort to be reasonable and mature. Similarly the 1966 TV season includes *Star Trek*, a program in which science fiction is treated seriously.

and must be slain by Bellerophon on his flying horse, Pegasus, is much like the monster that rises from twenty thousand fathoms in the black lagoon and must be slain by a movie hero and his flying aeroplane.

For adult science fiction, for real science fiction, it is to the magazines and the paperbacks (plus an occasional hardback) that one must turn. Even there, not all stories are "good." (But, then, come to think of it, why should anyone expect all of science fiction to be good, or even most of it? One of the best of the science fiction writers once said to an audience of avid science fiction fans, "Nine-tenths of science fiction is crud." The audience sat stunned and disbelieving, and then the writer added, solemnly, "Nine-tenths of everything is crud.")

Embedded in the crud, however, are stories that are entertaining, well-written and exciting—but, more than that, thought-provoking in an odd way that is duplicated in no other form of literature. Here you will find strange, new societies: some oriented primarily toward advertising and its psychology; some hidden in underground cities; some faced with the discovery of new, intelligent life forms; some faced with the depletion of resources or repletion of population; some in which telepathy and its implications are commonplace.

Is this important? Of course it is. Good science fiction is fun, but it also does something that no other form of literature does: it consistently considers the future.

We are living in a society which, for the first time, *must* consider the future. Until 1750, the average man was certain that, short of the Day of Judgment, the essential way of life would proceed much as it always did and always would, except for changes in the actual cast of characters playing out the human drama.

After 1750, more and more men became increasingly aware that society was changing in odd and unpredictable directions and would continue to do so; that what was good enough for the father would turn out to be not good enough (or, perhaps, too good) for the son; that as things had always been, they would not remain.

After 1945, men further became aware that even the mere

fact of continued existence of human society in any form was by no means to be assumed. The possibility of a new kind of Day of Judgment grew big.

Science fiction is based on the fact of social change. It accepts the fact of change. In a sense, it tries on various changes for size; it tries to penetrate the consequences of this change or that; and, in the form of a story, it presents the results to the view of the public, a public that needs more and more to have the possibilities of change pointed out to it before it is disastrously overwhelmed by it.

It is this which has always made it seem rather ironic to me that science fiction is continually lumped under the heading of "escape literature," and usually as the most extreme kind, in fact. Yet it does not escape into the "isn't" as most fiction does, of the "never was" as fantasy does, but into the "just possibly might be." It is an odd form of escape literature that worried its readers with atom bombs, overpopulation, bacterial warfare, trips to the moon, and other such phenomena decades before the rest of the world had to take up the problems. (Would that the rest of the world had listened sooner!)

No, no, if science fiction escapes, it is an escape into reality.

The writers of science fiction are themselves not always aware of what they are doing. Many of them might swear in all earnestness that they are interested only in turning out a craftsmanlike story and earning an honest dollar. To my mind, however, they represent the eyes of humanity turned, for the first time, outward in a blind and agonized contemplation of the exciting and dangerous future, not of this individual or that, but of the human race as a whole.

THE CULT OF IGNORANCE

Chapter 33   The Cult of Ignorance

On June 25, 1956, I watched *Producer's Showcase* on television and witnessed, in striking form, the conflict between the Need for Education and the Cult of Ignorance.

The Need for Education was brought home with the very first commercial, which pulled no punches. The sponsor, it seemed, needed missile engineers, and he set about luring such engineers to his Florida factory. He stressed the climate and the beaches, the good working conditions, the cheap and excellent housing, the munificent pay, the rapid advancement, the solid security. He did not even require experience. The effect was such that, I, myself, felt the impulse to run, not walk, to the nearest airport and board a plane for Florida.

Having overcome that impulse, and having brooded for half a moment on the shortage of engineers and technical men brought on by the ever-intensifying technological character of our civilization, I prepared to enjoy the play being presented, which was an adaptation of *Happy Birthday* by Anita Loos, starring Betty Field and Barry Nelson. I *did* enjoy it; it was an excellent play—but behold, the sponsor, who a moment before was on his knees, pleading for technically trained men, paid to have the following presented to his audience of millions.

Barry Nelson is a bank clerk who spends much of his free time in a bar because that is where one meets women (as he explains). The one setting is the bar itself and the cast of characters is a wonderfully picaresque group of disreputables with hearts of tarnished gold. Barry Nelson, in the course of the play, explains that he doesn't read books (he is talking to a librarian) although, he admits with seeming embarrassment, he once did. He explains that his father once paid him a sum of money to learn to recite the books of the Bible in order, and to show he can still do it, he rattles

This article appeared under the title of "The By-Product of Science Fiction" in *Chemical and Engineering News*, August 13, 1956.

them off, explaining that when he was younger he could recite them much more quickly. Thus, the audience is presented with an example of what book learning is, and it is clear to them that this sort of thing is useless and ridiculous and that Barry is wise to eschew books and confine himself to bars.

Betty Field, on the other hand, is a librarian; that is, an educated girl, since she implies, now and then, that she has read books. She is shy, corroded with unhappiness, and, of course, unnoticed by boys. In the play, she violates the teetotaling habits of a lifetime and takes a drink, then another, then another. Slowly, she is stripped of her inhibitions. The stigma of intelligence is removed, layer by layer, as she descends into a rococo alcoholism. The result is that the barflies, who earlier viewed her with deep suspicion, end by making a heroine of her; her alcoholic father, who beat her earlier, takes her to his heart; and best of all, the bank clerk, who had never noticed her earlier, makes violent love to her.

I repeat, I enjoyed it thoroughly. And yet, viewed in the sober gray light of the morning after, the play preached the great American stereotype which we might call the Cult of Ignorance. According to that stereotype, it is only in ignorance that happiness is to be found; and education is stuffy and leads to missing much of the happiness of life.

Is there some connection between this and the fact that the sponsor is having trouble finding technically trained men?

Yes, we need technicians. Society as a whole needs them or it will collapse under the weight of its own machines. But how are we trying to get them?

Is it sufficient for an industrial concern to lure missile engineers? What it amounts to is that engineers are being lured from one specialty into another, with the total number seriously short. If a community can get rich by taking in one another's washing, this sort of thing can work, but otherwise not.

Solutions have been suggested to the problem of the shortage of scientifically and technically trained men. Some advise that science teachers be paid more, that bright students be given scholarships, that industrial chemists and engineers devote time to teaching and so on. All these points are valuable, but do any of them go far enough? If one did, somehow, get a sufficiency of wonderfully expert science teachers, whom would they teach? A group of students, most of whom have been indoctrinated from childhood on with a thoroughgoing belief in the limitations of educated people and the worthiness of natural ignorance.

Think for yourself of the literary stereotypes of the "bad boy," the best of whom were Tom Sawyer and Penrod Schofield (with more modern examples populating radio and television). School is their enemy; schoolteachers hateful; book learning a bore and delusion. And who are the villains of the piece? The Sid Sawyers and Georgie Bassetts—little sneaks who wear clean clothes, speak correct English, and like school (loathsome creatures).

I have never stolen an apple from a neighbor's apple tree or rifled a watermelon from his watermelon patch (there being little or no opportunity to do so in the depths of Brooklyn), but even I was sufficiently seduced by the skillful wordcraft of the author to learn to detest the villainous teacher's pets who wouldn't engage in such lovable and manly little pranks, or who wouldn't play hookey and lie about it, or participate in a hundred and one other delightful bits of juvenile delinquency.

Perhaps it is our pioneer background, when school seemed merely a device to take a boy from his necessary chores and put him to work learning Latin verb declensions, to the thorough exasperation of his overworked father. Whatever it was, many of us can remember the scorn heaped by the newspapers on the "fuzzy-minded professors" of the Brain Trust of early New Deal days. And it is taken for granted that Adlai Stevenson was helped to his landslide defeats in 1952 and 1956 by his persistent revelation of intelligence.

Have you ever noticed the role played by spectacles in movies and television? Glasses in the popular visual arts of today are the symbol of developed intellect (presumably because of the belief on the part of the average man that educated people ruin their eyes through overindulgence in the pernicious and unhealthy habit of reading). Ordinarily, the hero and heroine of a movie or television play do not wear glasses. Occasionally, though, the hero is an architect or a chemist and must wear glasses to prove he has gone to college. In this case, he is constantly whipping them off at every forceful speech he makes, since you can't be virile and wear glasses at the same time. True, he puts them on to read a piece of print, but then off they shoot again, as he bunches his jaw muscles and assumes the more popular role of unpedantic valor.

An even better example is a Hollywood cliché that has been so efficiently ground to dust by overuse that even Hollywood dare not use it again (an almost incredible state of affairs). The cliché to which I refer is the one whereby it is assumed that a superbly beautiful actress, whom we shall call Laura Lovely, is ugly, provided she is wearing glasses.

This has happened over and over again. Laura Lovely is a librarian or a schoolteacher (the two feminine occupations that, by Hollywood convention, guarantee spinsterhood and unhappiness) and naturally she wears big, tortoiseshell glasses (the most intellectual type) to indicate the fact.

Now to any functional male in the audience, the sight of Laura Lovely in glasses evokes a reaction in no way different from the sight of her without glasses. Yet to the distorted view of the actor playing the hero of the film, Laura Lovely with glasses on is plain. At some point in the picture, a kindly female friend of Laura, who knows the facts of life, removes her glasses. It turns out, suddenly, that she can see perfectly well without them, and our hero falls passionately in love with the now-beautiful Laura and there is a perfectly glorious finale.

Is there a person alive so obtuse as not to see that (a) the presence of glasses in no way ruined Laura's looks and that our hero must be completely aware of that, and (b) that if Laura were wearing glasses for any sensible reason, removing them would cause her to kiss the wrong male since she probably would be unable to tell one face from another without them?

No, the glasses are not literally glasses. They are merely a symbol, a symbol of intelligence. The audience is taught two things: (a) Evidence of extensive education is a social hindrance and causes unhappiness; (b) Formal education is unnecessary, can be minimized at will, and the resulting limited intellectual development leads to happiness.

It is this stereotype of good human ignorance versus dry, unworldly education that we must somehow fight and conquer if we are ever to get sufficient quantity of raw material —that is, children who are brought up to respect and admire intelligence—upon which to apply the palliatives we suggest (money, security, prestige) to increase our supply of scientists and technicians.

What seems hopeful in this connection is that there is one entire branch of popular literature which is largely given over to the proposition that brains are respectable. That branch is known as science fiction (see Chapter 32).

Naturally, a science fiction story can be entirely frivolous, as, for instance, would be the case of a story dealing with a man who invents a device whereby he may unobtrusively see through walls and clothing. It should be obvious that, properly handled, a great deal of enjoyable ribaldry may result, but nothing much beyond that. A science fiction story can even be antiscience, as were a great many, several years ago, which described atom-shattered earths with scattered and primitive survivors, all yielding the pretty obvious moral that all this would not have happened if only men had avoiding poking their nose into science and had stayed close to the simple things of life.

But a significant fraction of science fiction stories have as their chief motivating force some kind of technical problem and as their chief characters, technically trained people.

I can cite some examples from among my own stories. One deals with a party of scientists who travel to a distant planet to find the reason for the mass death of an earlier colonizing party despite the planet's apparently ideal nature as a home for man. The answer turns out to be that the planet's crust is high in beryllium compounds and death is the result of insidious beryllium poisoning.

The second story deals with the efforts of a historian to gain permission to use the government's "time-viewing" machine in order to gain data on ancient Carthage. On the government's refusal, he engages the services of a physicist to build him a time-viewing machine of his own—with totally unexpected and tragic results.

In the first story, there is a consideration of the problem

of the expanding quantity of scientific data and the increasing realization of the inability of the human mind to cope with even a fraction of it. In the second, there is a description of what might take place in a society where government grants become the sole financial support of research. This sort of thing is, as you see, a cut above the low-budget Hollywood production of monster movies usually called "science fiction."

But both the story itself and the sociological background are, in a way, less important than the mere fact that although the individual scientist in such stories may be hero or villain (depending on whether he is intelligent and reader-sympathetic or intelligent and reader-unsympathetic), science and intelligence, themselves, as abstract forces, are represented sympathetically. Scientific research is presented, almost invariably, as an exciting and thrilling process; its usual ends as both good in themselves and good for mankind; its heroes as intelligent people to be admired and respected.

Naturally, science fiction writers do not deliberately go about doing this. If they did it deliberately, the chances are that their stories would play second fiddle to their propaganda and prove quite unpublishable; or if published, quite boring, and thus do more harm than good.

It merely happens that this sort of thing comes about almost unwittingly. However much a science fiction writer may think primarily of writing a good story and secondarily of making an honest living, he inevitably finds that every so often, he cannot escape making intelligence, education, and a scientific career attractive. That is the unavoidable by-product of science fiction.

Special Note: At the time this chapter was first published, it was greeted with the most profound lack of interest you can possibly imagine. A year later, the Soviet Union launched Sputnik I, the first satellite, and we suddenly found ourselves in a technological race with an opponent we had, until then, completely underestimated.

Suddenly, *everybody* was attacking the cult of ignorance, and things will never be quite the same again, perhaps.

However, I feel it only fair to point out that it is ordinarily desirable to see the cliff edge *before* you fall over and down. Screaming afterward is very easy.

Chapter 34   The Sword of Achilles

About 1200 B.C. (the story goes) the Greek forces were gathering in preparation for their assault on the city of Troy. An oracle had foretold that the assault would be in vain unless the young Achilles joined the Greek army. But Achilles' mother, the nymph Thetis, had dressed her son in women's clothes and had hidden him among the court ladies on the Aegean island of Scyros. She knew that if he went to Troy he would be killed there, and, motherlike, she found the prospect displeasing.

To Scyros came a delegation of Greeks under the leadership of the wily Odysseus. It would not have been polite to search the ladies in order to detect the man among them, but Odysseus specialized in the indirect method anyway. He laid out an assortment of fine clothing and jewels and asked the ladies to help themselves, which they did with much delight.

Among the luxuries, half hidden, was a sword. And one of the taller maidens strode forward, seized it, and flourished it with a shout. The "maiden" was, of course, Achilles, who thereupon went to Troy and his death.

Wars are different these days. Both in wars against human enemies and in wars against the forces of nature, the crucial warriors now are creative scientists.

Creative scientists are both born and made. The spark is there, presumably, to begin with, but it can all too easily be extinguished. A serious task facing educators today, therefore, is to devise methods of teaching that will foster creativity in youngsters.

But teaching for creativity is itself a wholesale consumer of creativity. It requires superlatively good teachers and highly imaginative techniques. To spread such education

This article first appeared in *Bulletin of the Atomic Scientists,* November 1963.

broadcast, even if it could be done, would be wasteful. Although all normal human beings possess a measure of creativity (when one thinks of the many discoveries that a child must make in the course of growing up, who can doubt that?) the gift is certainly greater in some than in others, and it may not always incline in the direction of the sciences. Clearly, then, if our society is to develop creativity in science with maximum efficiency, we must seek out the richest ore; we must find the children with the greatest potential and focus our best efforts upon them.

But how does one detect a potential creative scientist?

There are infant prodigies, of course. There could be no doubt that the young Arrhenius and the young Gauss were destined for great things if they lived, even if they had been miseducated. On the other hand, Isaac Newton showed no great promise until he was about sixteen. At a superficial glance it is even possible sometimes to confuse budding creativity with retarded mentality or with juvenile delinquency, both of which were suspected in the case of Thomas A. Edison.

Men have tried to devise tests for creativity, and they have sought to arrive at empirically selective criteria by listing the qualities that individuals known to be creative have in common. But all such tests and all such sets of criteria are alike in being fuzzy, uncertain, and extremely controversial.

What we need is a simple test, something as simple as the sword of Achilles. We want a measure that will serve, quickly and without ambiguity, to select the potentially creative from the general rank and file. We would not ask that such a test single out *all* youngsters with the spark we seek. I think most of us would be satisfied to cull out a subgroup in which the incidence of potential creativity in science was substantially increased over that in the general population.

I would like to suggest such a sword of Achilles. It is simply this: an interest in good science fiction. This suggestion is not a mere guess on my part. It is based upon estimates of (I believe) reasonable validity. Let me spell them out.

I am myself, among other things, a science fiction writer, and I know just how well my books sell. One of them has sold, in all American editions, including the paperback, some 400,000 copies. A number of these have been bought by libraries, where perhaps dozens of people have read each copy. On the other hand, many people may have bought a paperback copy and then merely leafed through it casually, without interest. Let us suppose, reasonably enough, that these two sources of error largely cancel each other; we can then place the total number of individuals interested in science fiction in the United States at 400,000.

This is a deliberately generous estimate, because I am told that my science fiction sells better than average and I have chosen the one of my books that has sold best. By this generous estimate, then, with the population of the United States totalling 180,000,000 (at the time this book was selling), we can say that one out of every 450 Americans is interested in science fiction.

Consider next that for a quarter of a century I have also lived and worked in the academic world and have moved in circles where I met many creative scientists. Half of these, I should say (and I do not refer to all the scientists I have known but only to those I judged creative), have read science fiction at some time in their lives.

At a recent conference on methods of teaching creativity in science which I attended, I suggested this estimate in private conversation, and the person to whom I was speaking maintained vehemently that not 50 percent but 95 percent of those present had some interest in science fiction. But let us put his higher percentage down to enthusiasm and stick to my 50 percent: one in two.

It could be argued that a scientist's interest in science fiction is merely a reflection of his professional preoccupation. I do not believe this is often the case, however, since one rarely begins reading science fiction in adulthood. The habit starts in adolescence, as a rule, and interest in science is stimulated by the reading rather than the reverse.

Compare, then, the conservative estimate of a one-in-two incidence of interest in science fiction among creative scientists with the generous estimate of one-in-450 incidence in the general population. One can only conclude that by the single, simple process of choosing all science fiction readers among, let us say, ten- to fifteen-years-olds, one can concentrate the incidence of potential scientific creativity by a large factor.

If there is any validity in this reasoning, and I am sure there is, it seems a shame that forces sometimes operate to inhibit a youngster's enjoyment of science fiction. English teachers often lump all science fiction into the group of unreadable material forbidden their students and will not accept, for instance, a book review of a science fiction novel as a reasonable offering in response to a homework assignment. (I have received innumerable letters from young readers complaining of just this.)

Many English teachers are not interested in either science or science fiction. Uncomfortable with tales of a world that is alien and seems fantastic to them, they take the easy way out and forbid the material. The tendency is diminishing, thank goodness, but I would like to see it disappear altogether.

Science fiction has its good examples as has every other branch of literature, and if English teachers, through lack of experience, have trouble distinguishing good science fiction from bad, they have only to ask the help (and I say this in all seriousness) of any bright twelve-year-old in their classes.

If science fiction were freely available in school libraries, I dare say our sword of Achilles would suit us very well. It would not, of course, pick up every potentially creative scientist. And the percentage of such promising persons in the population is probably so low that even after a manyfold concentration we would still be left with a subgroup containing a noncreative majority. Nevertheless, we would have rich pickings in the selected group, compared with those in an unselected lot.

And I defy anyone to suggest a better sword of Achilles.

Special Note: If anyone suspects that my estimates as to the importance of science fiction—as described in the previous three chapters—is so solemnly great that I am unable to see its lighter side, please be undeceived.

The final three chapters of this book represent a gentle satire of the field.

302

In a way, this, too, is the measure of my belief in the value of the field. I consider that value to be strong enough and worthy enough to withstand, without harm, a little fun at its expense.

Chapter 35   How Not to Build a Robot

As it happens, I have never been asked to serve as technical advisor for any television program. That is television's loss, of course, but I am a very busy man and have no time to feel sorry for TV. Let it suffer the consequences, say I.

I mean, of course, under ordinary conditions. In the 1964-65 season, however, I noted a tendency to go too far. I am referring to the program *My Living Doll.*

This series dealt with Robot AF-709, which had been secretly built by a roboticist at some space center. The roboticist was then assigned to Pakistan and he left the robot with his best friend, a psychiatrist named Dr. McDonald, for safekeeping. It was necessary for Dr. McDonald not to allow anyone to guess that the robot is a robot.

This presented some difficulties, for the roboticist had built the robot in humanoid shape. In fact, its name was Rhoda and it resembled a large woman of spectacular physique.

Isn't that disastrous? I could have prevented this. My infernal modesty prevents me from explaining in great detail at this point that I am an acknowledged world authority on robots—so I'll just mention the point briefly. I am an acknowledged world authority on robots.

If they had asked me I would have said, "But you *can't* build a robot in the form of a large woman of spectacular physique. That's poor robotic engineering."

The best kind of robot, of course, is one built out of metal, with a smooth, cylindrical body; a delicately tapering conical head; strong tubular limbs. There is a clanking somber

This article first appeared in *TV Guide,* January 16, 1965, under the title "Why I Wouldn't Have Done it This Way." Copyright © 1965 by Triangle Publications, Inc.

majesty about such a robot's geometry which few would exchange for the pitifully uneven outline of Rhoda the Robot. When I say few, I mean few roboticists.

The show made the point that the robot was built to test the effects of space environment upon astronauts. Therefore, it must be admitted, a plastic covering with properties close to those of human skin would be desirable, and the planes, curves, and joints of the human body should be imitated. *But*—and here is where the illogic of it all overwhelmed me—why imitate the *female* body when almost all astronauts are men?

Admittedly, the workings inside a robot are comparatively bulky, so the robot must be large. But if it were built in the male form, the necessary size would not be too remarkable. In the female form it becomes ungainly and attracts undesirable and deprecating attention.

Even with a taller-than-usual girl, there remains insufficient room for all the necessary internal equipment. There must therefore be marked irregularities and bulges with which to place controls for which there is no room in the torso proper. You can imagine the stares these bulges attract —of disapproval, I am sure.

All this has made it necessary for the producers to go to great trouble to find a large girl of spectacular physique, when they might just as easily have decided to look for a man of only slightly more than average size. That's a lot of trouble they've gone to for nothing.

Can they point out to me, or to any rational human being, what can possibly be gained by building the kind of imbalanced robot that is made necessary by the lopsidedness of the female frame?

I tell you, as I sat in my living room watching Rhoda the Robot standing there in a sheet (on the screen, of course) I felt strongly urged to examine her closely in order that I might estimate just how bad the imbalance was. Any roboticist would have felt that same urge.

Then there was the question of the robot's controls. Those of you who never watched the program won't believe it, but you have my word—there were exactly four pushbutton controls, designed to resemble moles, placed on the upper portion of the back.

They are not marked in any way, and Dr. McDonald, being not overly bright (most unusual, for psychiatrists, as a class, are noted for their high intelligence and keen perception, as one of them recently explained to me), could never remember which control did what. Furthermore, exposed as they were, the controls—particularly the on-off button—could be struck by accident.

It is laughable to suppose that four controls would be sufficient for a robot of the complexity of Rhoda. You have but to look at her with the trained eye of a roboticist to see that she would respond to more than four types of stimuli. Any child, then, would realize that a whole series of buttons would be necessary, plus several dials and toggle switches and a couple of adjustable screws. There is only one logical place for all the necessary controls, and that is upon the abdomen.

Consider the advantages to this location. In the first place, the robot's abdomen faces the operator, who is therefore ready, at all times, to manipulate the robot properly. He need not, as Dr. McDonald constantly had to do, lift or turn Rhoda to get at her back.

Secondly, whereas the back, thanks to the fashions in feminine clothing, is exposed to accidental touch, the abdomen is always covered by several layers of textile. Abdominal controls would thus be better protected against unauthorized tampering or casual fingering.

Most of all the abdomen supplies a perfect place for protective positioning of the on-off button. For security, it can be placed within the umbilicus.

After all, I consider it highly significant that the robot abdomen was never exposed throughout the duration of the show. All through the first segment and through part of the second, the robot wore nothing but a sheet that covered her from armpits to mid-thigh. To the trained mind, this is significant.

What, after all, could the robot be hiding but her abdomen? She was protecting her controls! On two occasions, once in the first show and once in the second, the robot made as though to remove the sheet and on both occasions, Dr. McDonald stopped her in great agitation.

This was terribly disappointing to me, for had those controls been exposed, it would have proved to me that the producers had received good practical advice after all and that the so-called "controls" on the robot's back were merely a blind to conceal the truth for security reasons.

With this in mind, and filled with a pure scientific curiosity, I distinctly remember rising to my feet on each occasion when the robot seemed about to remove the sheet, crying, "Don't stop her, you silly ass!"

But what can you expect of a man like Dr. McDonald? He did stop her, and I suspect that the reason the show did not last for more than one season was that this highly expensive and intricate robot was ruined for lack of proper handling.

And that is another thing. Dr. McDonald was a poor choice as a guardian for the robot. A roboticist such as myself would have seen Rhoda for what she was—a poorly designed robot who had to receive very careful treatment and very sensitive handling if she were to flourish.

Consider that a psychiatrist must sit all day and be immersed in the Freudian problems of his female patients. Is there any wonder that he finds himself bashful in the presence of women? And as a bachelor who, of necessity, can have had little experience with women, Dr. McDonald would have to be phenomenally shy and modest. How could he be expected to handle Rhoda with the necessary skill and authority?

Then, too, Dr. McDonald seemed incapable of understanding the simplest aspects of robotic engineering. For instance, on several occasions, the robot made the perfectly simply robotic statement, "I do whatever I am told to do."

Well, of course! And yet, as she stood there in her sheet making the statement, Dr. McDonald blanched and seemed perturbed.

But why? A robot must do whatever it is told to do, as long as that is consistent with the circuits impressed upon her brain and consonant with the purposes for which she was designed. Any child would know that.

Rhoda the Robot would do anything, simply anything in the world, that would help a hard-working roboticist figure out the effect of space environment on the human body. That would be its great contribution to science and humanity.

THE INSIDIOUS UNCLE MARTIN

What is there about this logical situation, then, that could have upset Dr. McDonald? When the poor robot, badly designed to simulate a large woman of spectacular physique, offered to perform its natural robotic duties by doing exactly as it was told, what thoughts can have passed through Dr. McDonald's lay mind?

I suppose no one will ever know.*

Chapter 36 The Insidious Uncle Martin

A Martian on television?

When the rumor of this reached me some years ago, I could scarcely believe my ears. I could think of nothing more exciting, more thrilling, more scientifically useful than to put a Martian on television. I waited eagerly, therefore, for the program called *My Favorite Martian*.

For me, this was all particularly significant. For over a quarter-century, you see, I have read and written science fiction stories, so that I am thoroughly familiar with the appearances of Martians as worked out by some of the best minds in America (including my own—see Chapter 23).

While I counted the hours until the first showing of the program, I went over some of the descriptions of Martians I had seen in science fiction stories. There were, for instance, vaguely humanoid Martians which, however, were tall and spindly, with thin limbs and bulbous chests. There were Martians with faces like chrysanthemums; Martians with tentacles like octopi; Martians that resembled frowzy ostriches; Martians reminiscent of large, feathered worms.

---

* I suppose people *will* know, however, that I am kidding. I greatly enjoyed *My Living Doll*, was sorry to have it leave the air, and, most of all, I yield to no one—to no one, I tell you—in my pure and scientific admiration for the construction-design of Miss Julie Newmar, who played Rhoda the Robot to perfection.

This article first appeared in *TV Guide*, March 5, 1966, under the title of "Can You Spot the Family Resemblance?" Copyright © 1966 by Triangle Publications, Inc.

Of course, Martians were occasionally pictured as beautiful women, equipped with a maximum of charm and a minimum of clothing, but I never took that seriously. You have only to think about the matter rationally. It's very cold on Mars; beautiful Martian princesses would have to wear expensive fur coats and where could you get expensive fur coats on Mars?

It's the consideration of these little points that distinguishes the thoughtful and talented science fiction writer from the amateur bungler.

But never mind all that. The real thing, the true Martian, would now be clearly visible on the television screen. All doubts would soon be put to rest.

With beating heart and panting breath, I watched—and there, into my view, came the creature whom all the world was soon to know as Uncle Martin.

I started from my chair in amazement! Was it possible? The being looked very much like an earthman.

Of course I am not easily fooled. My science-fictional training makes me a keen-eyed sleuth with respect to extra-terrestrial details. I watched for small deviations from the norm that the average American would not have noticed in a thousand years.

I kept my eyes peeled for the presence of six or seven fingers on each hand (or even on just one), or for an extra thumb. I watched shrewdly for a second head that might be concealed in one of the creature's pockets; or for a tail that might flick occasionally down a pants leg and out onto the floor. Little things like that, which would mean nothing to an ordinary viewer, would have been highly significant to me.

But I observed nothing at all, until my young son pointed out that Uncle Martin had a pair of antennae that appeared now and then. I had missed those.

Antennae, eh? Was this decisive? I began to watch the earthmen about me in the streets and after several days, I felt convinced. Earthmen did not have antennae; at least not those in the Boston area. A good point!

A lesser man might have allowed this to convince him. On the strength of the antennae alone, he might have concluded that Uncle Martin, despite his earthmanlike appearance, was a Martian. However, these television people would have to stay up all night to put one across an experienced science-fictioneer such as myself.

I wanted more evidence. I spent weeks plunged in thought; neglecting my work while I pondered the problem. Uncle Martin seemed so at home on earth, and yet surely he found conditions quite strange here. It's cold on Mars, for instance, and very dry. Doesn't Uncle Martin find it extraordinarily hot and damp here on earth? If so, he showed no signs of it.

Of course, it was clearly demonstrated in one program that Uncle Martin had a body temperature high above that of earthmen. Could this mean he was unaffected by either Martian cold or earth's heat? Efficient perspiration glands could take care of the dampness. This point, then, was not conclusive either way. Tirelessly, remorselessly, I passed on to other considerations!

What about the atmosphere? The Martian atmosphere is less than a hundredth as dense as our own and contains no oxygen. Oxygen is an active chemical and would undoubtedly be poisonous to a Martian who was unaccustomed to its presence. The question, then, was how Uncle Martin managed to stay alive while breathing our air.

But was he breathing our air? I did not dare jump to conclusions. Carefully, I watched program after program, trying to detect that rhythmic, tell-tale rise and fall of the chest. Unfortunately, I could not be certain whether I detected such a motion or not.

Since it is scientifically important, in such matters, to run a control, I chose another character in the weekly drama to see if the chest of an obvious earthling would rise and fall visibly. On a purely random basis, I chose the attractive landlady, and observed her prominent chest intently on her every appearance for five or six programs. By that time, there was no question of it. Her chest rose and fell as she breathed, but I still remained uncertain in Uncle Martin's case.

Again, the evidence remained inconclusive.

And then came the solution. Gravity! Of course! Martian surface gravity is only two-fifths as strong as that of the earth. Any creature adapted to Mars would find itself weighed down on earth. It would walk only with difficulty;

raise itself only after obvious exertion. Life on earth would be a constant torture for it.

Yet Uncle Martin showed no difficulty in moving. He seemed rather light-footed and graceful, in fact. I checked my control at once, staring, with scientific fervor, at every movement of the landlady. The reaction to earthly gravity seemed the same in both cases.

At last, I had a conclusive point. My careful analysis of the situation had taken me years, but it was worth it. The conclusion I reached will, I am sure, rock the nation.

That conclusion is simply this: *Uncle Martin is not a Martian!* He is an earthman, nothing more and nothing less.

Yet he is not merely an earthman, either. He *does* have antennae; I have seen them myself. He has all kinds of Martian powers, too. He can make himself invisible, for instance, and he can make objects move by pointing his finger at them.

Of course you will suspect that these powers may be faked. I suspected that, too, but it's pretty difficult to fool an experienced science fiction writer. I know all the tricks.

It could be a matter of distraction, for instance. You could be watching Uncle Martin's finger move in the direction of a chair, for instance, while someone else rushed in, unobserved, and actually moved the chair. Or perhaps Uncle Martin had a long tube attached to his finger, a tube colored a neutral gray so that it would be unnoticeable. And when it seemed Uncle Martin had disappeared, it might be that someone had actually moved a screen in front of him, one that blended with the background.

I thought of a dozen extremely subtle dodges like that, but convinced myself that not one of them was used.

We are left, therefore, with an Uncle Martin who is clearly an earthman but who, as clearly, possesses undeniable Martian powers.

We can only conclude, then, that the Martians are indeed behind the program. And, undoubtedly, behind similar programs that will succeed it. But why? If the Martians are going to demonstrate Martian powers, why not use a real Martian? Why use an earthman?

My young son inadvertently gave me the necessary clue;

which my giant brain seized on at once. He said, recently, "Gee, I like Uncle Martin."

Of course!!! Would my son like a Martian if it were presented to him in its real guise as a feathered worm or a scaly octopus? Never! It is clear that the Martians are deliberately presenting a false image to the world! They are subverting our youth! They are shrewdly winning our hearts!

We are being presented a "Martian" who is made to seem a creature just like ourselves; one who has un-earthly powers but who uses them in kindly fasion to help the young man he lives with, and to keep the landlady out of trouble. Even the detective, who constantly suspects that Uncle Martin is not what he seems, is treated with gentleness.

The conclusion is that the Martians are eager to help us and that they even love their enemies. At least that is the conclusion they *want* us to come to.

But is such a conclusion valid? If it really were, then why this elaborate flummery, so cleverly concealed that it took all of even my ingenuity years to see through it?

Can it be that after we have been properly brain-washed into becoming soft on Martians, these creatures will show themselves in their true colors, and take over?

It would be naive and fuzzy-minded to think anything else!

Earthmen, awake! You can't do business with Martians! Quickly, before it's too late! Open your eyes to the vicious Martian conspiracy all about us! Don't be fooled by the insidious Uncle Martin!

If we act now, we may save earth, but time is running out. Act *now!*

Chapter 37   *The Lovely Lost Landscapes of Luna*

Some time ago, a planetary probe yclept Mariner IV passed in the neighborhood of Mars and violated the chasteness of our sister planet with a series of twenty-one photographs. The veil of distance was ruthlessly torn away and the scars of Mars were brought plainly out into the open.

No canals! Just pockmarks like those of the moon. One crater was 75 miles across.

The last vision of an exotic world slipped into limbo and the solar system grew one additional stage lonelier.

I am a science fiction writer of the present, but my youth dates back to the early 1930s when I was merely a science fiction reader, heir to a romance forever gone. In my days, you see, the solar system was peopled with mysterious races, gorgeous princesses, lavishly preposterous beasts and monsters, plants of all sorts from deadly to intelligent.

It was a solar system the like of which we shall never see again—and it was science that killed it.

Until modern times, men believed only the earth was inhabited. Average men, that is. To sophisticates, however, even in ancient times, it seemed clear that the moon and the sun were worlds and that the planets might be worlds, too. And, until modern times, an uninhabited world was a contradiction in terms. What use was a world unless it was inhabited by creatures very much like ourselves? A world uninhabited was a world wasted and such inefficiency was a slur on either God (if one were religious) or on the logical machinery of the universe (if one were not).

Thus, the Greek satirist, Lucian of Samosata, who lived in the 2nd century, A.D., wrote of an earthman who visited the moon, which he found to be inhabited by people who were at war with the people of the sun over the right to colonize Venus.

Even as late as 1800, the authentically great German-English astronomer, William Herschel, populated the solar system. He thought that the sunspots were breaks in the sun's fiery atmosphere through which one could see the dark inner surface of the sun itself—a sun which might be cool, and even inhabited.

In 1901, H. G. Wells in *The First Men in the Moon* was still carpeting the moon with plants. He went on to describe intelligent moonmen, living underground.

Why underground? Reality had begun to obtrude itself. As soon as telescopic observation of the moon became a possibility in the 17th century, it became obvious that the moon had neither air nor water on it. There were flat, dark tracts of the lunar surface that were named "seas"; seas with beautiful names: "The Sea of Tranquility," "The Sea of Serenity," "The Sea of Dreams."

But, alas, if they were serene and tranquil it was because no breath of air existed to stir them. If they were marked by dreams, they were sad dreams of an inhabited world that wasn't, a dainty vision of a smaller, more delicately built world than our own. The dream was replaced by the nightmare of seas of dust, ever-silent and ever-unchanging crags, a relentless slow-moving sun, a relentless freezing-creeping night. Modern science added the never-ending rain of deadly radiation.

Science fiction writers might still write of a populated moon despite all this, but the heart had gone out of it. To fly directly in the face of science gave science fiction a bad name, and, as writers grew more conscientious, the good name of science fiction grew to be more valued.

Of course, there was the other side of the moon, the part we never see from earth. What if the moon were egg-shaped, bulging toward us? Earth's gravity would seize upon this bulge and halt the moon's rotation, freezing it in place with that bulge facing us. We would see what was, in effect, a huge airless mountain. On the other side, though, in the moon's rich lowlands, would be air and water, and a population. It was a beautiful idea and there was no way to disprove it, since the other side of the moon was forever invisible to us on earth's surface.

And then, in 1959, the Russians sent Lunik III around the moon, and peeked at the other side. Gone were the seas, the air, the fleecy clouds; lost was the lovely landscape of Luna. The other side of the moon, it turned out, was worse than the side that faced us, more mountainous and just as bleak.

Well, then, underground? Like the moonmen of H. G. Wells? No. Scientists have considered the matter, and they advance all sorts of reasons to suppose that, *at most,* there may be some bacteria or equally simple life under the moon's skies. No more.

And, of course, let's not even mention the sun. Its outside is 10,000° F. and the sunspots, despite Herschel, only look black in comparison. They are about 7000° F. at least. What's more, there's no cool interior to the sun. It gets hotter all the way to the center, which stands at 25,000,000° F.

Yet in the 30s it wasn't the moon (and certainly not the sun) that was the heyday of life. We knew the worst about the sun and suspected the worst about the moon.

But there was Mars! In the case of Mars, science was on our side!

After all, had not the Italian astronomer, Giovanni V. Schiaparelli, discovered canals on Mars in 1877? Did not legitimate astronomers, such as the Frenchman, Camille Flammarion, and the American, Percival Lowell, insist that such canals could only be built by intelligent beings, and did that not mean that Mars was inhabited by people?

How many science fiction stories centered on Mars! How many lovely princesses in shining metal brassière cups, some flowing transparent drapery, and very little else, sat astride six-legged steeds, while stalwart earthmen fought on their behalf with gigantic swords!

Of course, the reasoning went, Mars was a smaller world than earth and had cooled sooner, thus getting an earlier start. Its civilization was far more advanced than ours and also far more decadent. Water was slowly disappearing, and the canals had been built as a desperate effort to hold off the end. The weary old Martians might be facing the inevitable with philosophic equanimity, offering their teachings to the brash youngsters of earth. Or else, turned evil out of need, they planned to invade the rich, young planet next toward the sun, killing or enslaving the native population (us!).

How often I thrilled to the machinations of the evil geniuses of Mars and their inevitable frustration by gallant earthmen.

Or perhaps the Martian civilization had already gone and earthmen had come and built upon its ruins. The small sun of Mars shone out of a purple, cloudless sky on what was left of the muddy canals, while human archaeologists pored endlessly over the remains of the mysterious dead Martians.

Of course, disquieting news kept percolating out of the observatories. The atmosphere of Mars was as thin as the air on top of Mount Everest, or thinner. There was practically no oxygen in what air there was. Many astronomers couldn't see the canals; and besides there was very little water on Mars.

We fought them off. We clung to Mars. It was our best hope. They couldn't take it away from us.

But they did. Oh, they seemed to give in here and there. There definitely *was* water on Mars; the ice caps that were clearly visible were indeed frozen water (and not solid carbon dioxide or anything heartbreaking like that) but there wasn't much. And the greenish areas on Mars might indeed be plant life—but not jungles, or trees, or even grass. Just, at the most, primitive lichen-like vegetation.

Then came Mariner IV and out the window went the canals altogether. No sign of them. Those astronomers that did see canals were probably seeing irregular lines of craters just at the limits of visibility, and the mind drew in straight lines where none existed.

Furthermore, the existence of craters not only showed how little air and water there was, but indicated that the air and water had been sparse indeed for long millions of years.

Princesses? I'll settle for lichens.

How about Venus? It is closer to the sun, took longer to cool down (according to the 1930-ish thoughts of science fiction writers), and so was younger than earth. In fact, it was obviously a younger world, because its atmosphere was filled with clouds. It had to be a dank, drizzly world, filled with a rankly growing jungle far more brilliantly riotous than anything the earth had ever seen.

Stories were written of an environment on Venus in which mold threatened everything, in which rapacious plants fought a riotous civil war without pause or quarter. What's more, it was thought, Venus kept one side eternally to the sun, though the cloud cover kept the temperature from becoming unbearably high. The dark side of Venus, with its eternal light, had a completely different environment, murkily mysterious, with warm air sweeping in from the day side and freezing into mountains of solid oxygen and nitrogen.

Or were the eternal clouds of Venus the signs of an enormous ocean on its surface? So enamored was I of this possibility that in 1954 I wrote a novel about the planet, depicting it as consisting of one large ocean spreading over the entire planet. I peopled that ocean with the weirdest creatures I could imagine, including a large octopuslike thing a mile across.

The best of it, you see, was that we were safe. It was impossible to look beneath the cloud cover of Venus. We could have this world exactly as we wanted it, and without scientific back talk.

The astronomers, however, played cat-and-mouse with the cloud cover itself. It was formaldehyde, theorized one. Gasoline, said a second. Dust, said a third. We were all on tenterhooks, before they finally settled on water after all.

Then, of course, the astronomers decided there was no oxygen in the atmosphere (astronomers *never* find oxygen anywhere outside earth, see Chapter 13). Science fiction writers argued themselves out of taking that seriously. After all scientists could only see the air *above* the clouds, what did they know about what was below?

But then astronomers learned to pick up radio signals from some of the planets, including Venus. The kind of radio signals Venus was sending out could only be sent out by a pretty hot object, once that was about 600° F. in fact. In 1962, the Venus-probe, Mariner II, was sent out and that temperature was confirmed. Venus was *hot*.

Venus was indeed covered with an ocean, as I had predicted in 1954. The only trouble was that the ocean was all in the form of steam. That cloud cover did not indicate that Venus had an ample water supply; it was Venus's *entire* water supply.

What's more, it turns out that Venus actually rotates slowly with respect to the sun. There is no perpetual night side, and no refuge from the heat. Venus is *all* hot.

Exit Venus. Exit the most beautiful and the most deadly jungle of the solar system; exit the largest ocean.

There wasn't much hope for Mercury ever. It was too close to the sun, and it kept one side to the sun at all times and one side away. But what about the "twilight zone" in between? Air could be piped in from the frozen oxygen mountains on the night side.

No go! The astronomers explained that in detail. The orbit

of Mercury is quite elliptical. Each revolution, it moves in close to the sun, speeding its flight, then moves away from the sun, slowing down. As a result, its surface swings like a pendulum so that each side of the so-called twilight zone gets 44 days of sun and 44 days of night. There just isn't any twilight zone.

In 1965, things got worse. It turned out, according to the radar beams bounced off Mercury, that the planet did turn slowly after all. There was no night side, either. Every part of the planet had long periods of sunshine. There was no frozen oxygen anywhere.

Beyond Mars are the giant worlds of the solar system—Jupiter, Saturn, Uranus, and Neptune—sharing among them twenty-nine satellites, five of them quite huge.

In the Thumping Thirties, we peopled them all. Stories were written that were laid even on giant Jupiter and Saturn. Some viewed Saturn as a world of prairies, a gigantic Wild West with enormous herds of cattle. And this was something, for Saturn has a surface area some 80 times as large as that of the earth, if what we see as the surface is really the surface.

As for the satellites— In my own stories of the late 30s, I had menaces on Ganymede and Callisto, two of Jupiter's moons. Titan, the largest moon of Saturn, was another favorite.

Nor were mere distances of the planet from the sun a stopper. One of the great stories of 1930 pictured a solar system facing doom as a result of the machinations of the evil inhabitants of Neptune.

It was a losing fight, however. The outer planets are too cold; their atmospheres are too deep and thick; and besides, the nosy parkers have analyzed it and those atmospheres are irrevocably poisonous. As for the satellites, there is only one on which an atmosphere has definitely been located and that one is Titan. It has a thin blanket of air, but—you guessed it—it's poisonous.

Half-heartedly, some astronomers speculate that the temperature on Jupiter might be higher than we think, might even be high enough to be comfortable—if you like breathing poison. Besides, they now think that the outer planets are almost entirely hydrogen—hydrogen gas in the atmosphere; hydrogen liquid below; metallic solid hydrogen at the center.

What else is there? Comets?

Jules Verne in one of his novels had a comet strike the earth a glancing blow and carry off some earthmen, who live upon the comet, more or less comfortably, for a considerable length of time. There was even an ocean on the comet.

But comets, we now know, are merely enormous volumes of gas and dust surrounding an asteroid-sized object that is a mass of pebbles held together by frozen gases.

Science has not even had the grace to keep its fingers off earth itself. Little by little, as explorations proceeded, the lost tribes and hidden civilizations vanished. Atlantis departed for good; darkest Africa yielded no "She"; and highest Tibet no "Shangri-La." No one has discovered anything but miserable primitives in the dank vastness of Amazonia, and the enormous southern continent the ancients dreamed of shrank into a desert Australia peopled by aborigines and an ice-bound Antarctica peopled by nobody.

Across the last horizon of earth, romance fades wistfully.

And earth's interior? What lies there? Interest in the bottomless caverns within the earth dates back to *The Iliad* at least. No mythical Greek hero was any good at all if he didn't invade the Underworld. Theseus did it. Hercules did it. Odysseus did it. The Romans copied and let Aeneas do the same.

Modern writers had their Underground as well; and stories were written about worlds within our own, about hollows with a radioactive "sun" at earth's very center. There were oceans and continents in the interior, monsters and men.

But even that innocent sport was taken away. By measuring the earth's density, by studying earthquake waves, and in a dozen other meddlesome ways, geologists have become quite convinced that the earth is absolutely solid. It is not hollow, and it cannot even have caves that penetrate more than a couple of miles or so.

Where, then, does that leave us? Nowhere in the solar

system, except on earth's surface, has any place been made safe for humanity. Nowhere may we expect to find cousins, kindly mentors, dangerous enemies. We are alone!

Well, not quite alone. There are other stars, with other families of planets (see Chapter 22). But they are far, far away; hard, hard to reach; and reserved, apparently, for a distant, distant time (see Chapter 31).

No, no, the stars are not enough. It's the solar system we want, the solar system they took away from us thirty years ago.

The solar system we can never have again.